I0819474

LEGENDS OF WARFARE

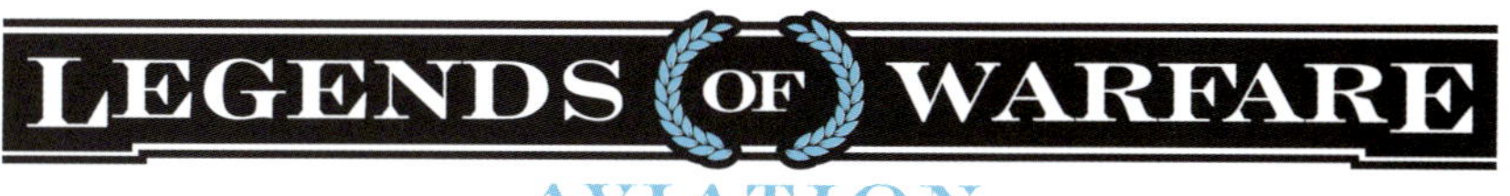

AVIATION

SBD Dauntless

Douglas's US Navy and Marine Corps Dive-Bomber in World War II

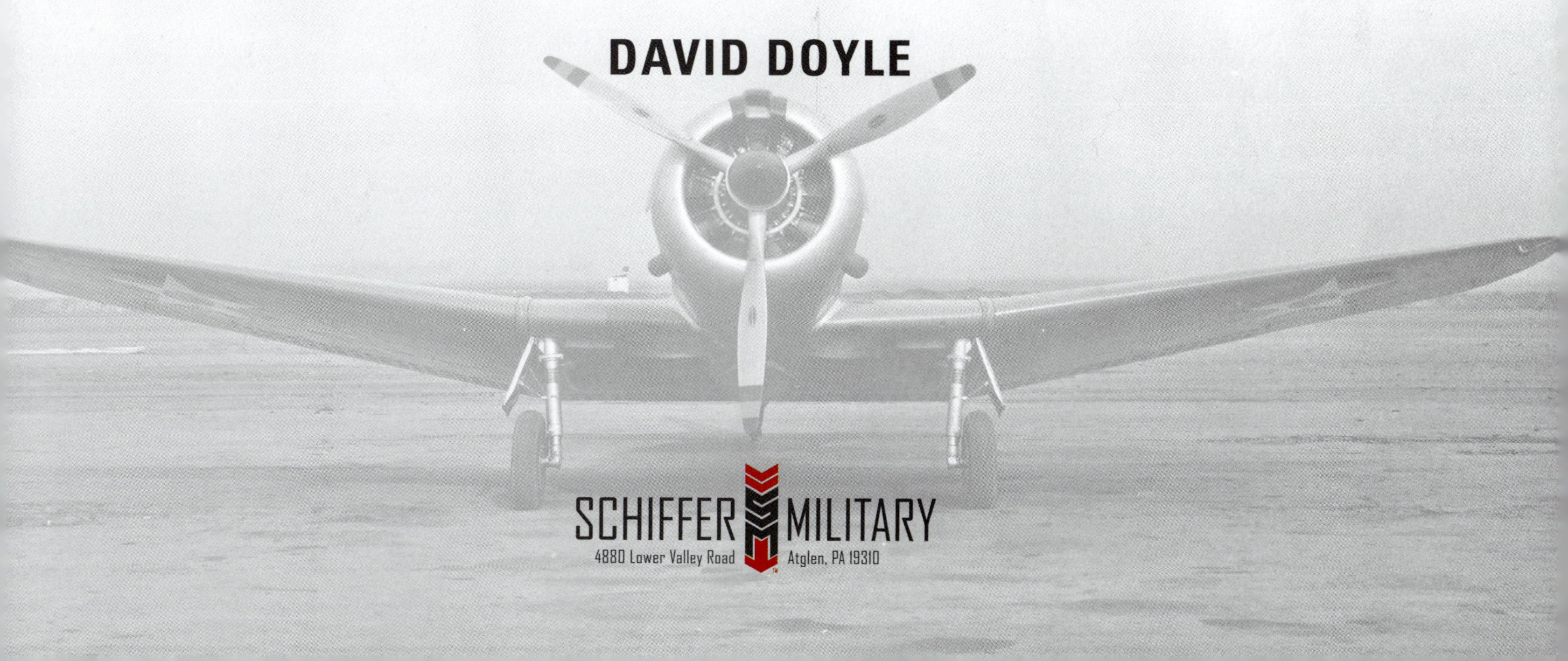

DAVID DOYLE

SCHIFFER MILITARY
4880 Lower Valley Road Atglen, PA 19310

Library of Congress Control Number: 2019934858

Designed by Justin Watkinson
Type set in Impact/Minion Pro/Univers LT Std
Front cover photo by Rich Kolasa

ISBN: 978-0-7643-5846-3
Printed in China

Published by Schiffer Publishing, Ltd.
4880 Lower Valley Road
Atglen, PA 19310
Phone: (610) 593-1777; Fax: (610) 593-2002
E-mail: Info@schifferbooks.com
www.schifferbooks.com

Acknowledgments

Creating this book has been very much a team effort and could not have been done without a great deal of help from many of my friends. Among those contributing to this effort are Dana Bell, Tom Kailbourn, Rich Kolasa, Scott Taylor, and Stan Piet, as well as the staffs of the National Museum of Naval Aviation, the Naval Historical Center, the American Aviation Historical Society, the National Archives and Records Administration, and the National Museum of the United States Air Force. Through all of this, the Lord has blessed me with a wonderful and supportive wife, without whose encouraging words (and scanning!) none of this would be possible. Thank you, Denise!

Contents

Introduction

Dive-bombing had its origins in World War I, when it was found that descending toward the target produced greater accuracy than horizontal bombing produced at that time. During the inter-war period the angle of attack increased, until nearly vertical. Dive-bombing reached its zenith during World War II. During the war, considerable work went into developing more-precise methods of delivering munitions, including air-launched rockets. While the SBD Dauntless was not the final dive-bomber to see fleet service with the US Navy, that honor going to the SB2C Helldiver, the Dauntless remains the most famous.

The SBD is a direct descendant of the Northrop BT-1. In 1934, when the BT-1 was designed, Donald Douglas owned 51 percent of Northrop Corporation. Leading the BT-1 project was designer Ed Heinemann, who later rose to fame as the designer of the Douglas A-20 and A-26 attack aircraft, the Douglas AD-1 Skyraider, and of course the SBD.

While the BT-1 was considered innovative, there was still room for improvement; thus production ceased after only fifty-four examples were built. It was to be superseded by an improved model, the XBT-2.

However, before the BT-2 could enter production, Douglas bought the remaining 49 percent of Northrop Corporation. Jack Northrop, on the other hand, would not be long absent from the aviation industry, quickly establishing Northrop Aircraft.

Because the Navy incorporated a code for the manufacturer's name in Navy aircraft model numbers, the XBT-2 became the XSBD-1—eXperimental Scout Bomber, Douglas.

Thus began the production of the aircraft that would come to be known as the Dauntless, or, in US Army service, the Banshee. The Dauntless would be present at Coral Sea, at Midway, off North Africa in Operation Torch, and in countless other actions, primarily in the Pacific. It remained in service through VJ-Day.

Northrop Corporation, of El Segundo, California, designed and built the XBT-1 prototype in response to a 1934 US Navy requirement for an all-metal, monoplane scout/dive-bomber aircraft. Notwithstanding the name of the manufacturer and some design differences, the XBT-1 and its production version, the Northrop BT-1, would in time prove to be direct ancestors of the Douglas SBD Dauntless dive-bomber. As seen on this BT-1 parked on a ramp at Northrop's El Segundo, California, plant, the large air scoop on the bottom of the fuselage aft of the cowling was a feature that differentiated the BT-1 from the BT-2, which had a much-smaller scoop in that position. *Naval History and Heritage Command*

Except for the large fairings under the wings, which housed the rearward-retracting main landing gear, and the two-bladed propeller, the Northrop BT-1 was similar in appearance to the soon-to-be Douglas Dauntless. This example is US Navy Bureau Number (BuNo) 0606, with markings for Bombing Squadron 5 (VB-5). A total of fifty-four BT-1s were produced. They were powered by the Pratt & Whitney R-1535-94 engine, rated at 825 horsepower. *Naval History and Heritage Command*

The XBT-2 prototype, BuNo 0627, introduced several improvements over the BT-1, including a three-bladed propeller, Wright XR-1820-32 radial engine, and inward-retracting main landing gear, with the bulky landing-gear fairings no longer necessary. When Northrop co-owner Donald Douglas became the majority shareholder in Northrop Corporation in 1937, he dissolved the Northrop Corporation (which later would be reincorporated), and the Northrop plant became Douglas Aircraft's El Segundo Division. Henceforth, production of the aircraft known as the Dauntless dive-bomber would be carried out by Douglas Aircraft. *National Archives*

CHAPTER 1

SBD-1

Production of the new dive-bomber began in El Segundo in June 1940. Following the then corporate custom of naming Douglas aircraft with words beginning in "D," the SBD was given the moniker "Dauntless" in October 1941. Orders for the new dive-bomber totaled 144 aircraft; however, only fifty-seven of these would be completed as SBD-1s (Bureau Numbers [BuNos] 1596 through 1631—the XSBD-1 prototype had been 0627). The rest of the order would instead be completed as the improved SBD-2.

The primary user of the SBD-1 would be the US Marine Corps, as a result of the Navy feeling that the SBD-1 was not combat ready. Among the reasons for this assessment was the lack of armor to protect the crew, as well as the lack of armor protection for the fuel tanks, which were not self-sealing. Those fuel tanks, incidentally, held only 210 gallons of aviation gasoline, which limited the aircraft's maximum range to just under 900 miles. Under combat conditions burdened with 1,200 pounds of bombs—1,000 pounds on the centerline, forward-swinging bomb crutch and 100 pounds under each wing—this resulted in a combat radius of a meager 200 nautical miles.

Marine Bombing Squadron 2 (VMB-2) was the first unit to be assigned the Dauntless, followed in quick succession by VMB-1. VMB-1 was redesignated VMSB-132 in July 1941 and the following winter was deployed to Guadalcanal in October 1942.

Once Douglas Aircraft took over the El Segundo plant, the company redesignated the XBT-2 the XSBD-1, and it served as the prototype for the first production series of Dauntlesses: the SBD-1 and SBD-2. A total of fifty-seven SBD-1s were completed, all of which were delivered to the US Marine Corps, with acceptances beginning in June 1940. As seen in this May 14, 1940, photo of SBD-1 BuNo 1596 with its main landing-gear doors removed, a key identifying feature of this model was the large carburetor air-intake scoop on the top of the cowling. *National Archives*

Douglas SBD-1 BuNo 1596 is marked with the USMC emblem on the fuselage below the rear of the aft canopy. A bomb is mounted in a shallow recess under the fuselage; a bomb crutch, designed to swing the bomb out to clear the propeller during a dive-bombing run, is attached to the sides of the bomb. The bomb crutch also was referred to as the bomb displacement gear, or bomb yoke. The Dauntless's "SBD" designation would lead to its nickname, "Slow but Deadly." *National Archives*

The stripes on the rudder of SBD-1 BuNo 1596 were painted, fore to aft, dark blue, white, and red. The small tailwheel was designed for carrier-based application. SBD-1s were assigned BuNos 1596–1631 and 1735–1755. These planes had no armor for the crew or self-sealing fuel tanks and were not considered combat ready. *National Archives*

The left side of the midsection of the fuselage of SBD-1 BuNo 1596 is displayed close-up in a photo taken May 14, 1940. Visible inside the canopy is the pilot's rollover frame. To the far right is the door to a storage compartment for an inflatable life raft and emergency rations. The USMC emblem appears to have been printed in colors. *National Archives*

In a companion view to the preceding photo, the sliding canopy sections of SBD-1 BuNo 1596 are open. Aside from the windscreen assembly, there were four sections to the canopy. The second section from the front was stationary. The pilot's canopy slid aft, and the two sections above the observer/gunner's cockpit slid forward. *National Archives*

The observer/gunner in the SBD-1 sat inside the ring mount for a .30-caliber machine gun, as seen in this view facing aft. When not in use, the gun was trained aft and the barrel was stored in a recess called the gun tunnel, equipped with two doors in the top of the fuselage. Below the receiver of the machine gun are racks for ammunition boxes. *National Archives*

A 1,000-pound bomb installation is illustrated on the centerline rack of SBD-1 BuNo 1596 on May 14, 1940. The front of the yoke-shaped crutch was designed to pivot on the short post projecting through the aluminum skin on the bottom of the fuselage. Bomb shackles held the bomb in place until released. *National Archives*

There were provisions for mounting a rack for a bomb up to 325 pounds under each wing of the SBD-1, as seen in this view of a 100-pound bomb installation under the right wing. *National Archives*

In this prewar photograph, Douglas SBD-1 BuNo 1603 is displayed in the markings of the commander of VMB-1, with the Marine Corps emblem visible on the fuselage below the rear of the canopy. Beyond this plane is SBD-1 BuNo 1605. The prewar national insignia, a white star with a red circle at the center, on a blue circle, was on the top and bottom of each wing, but not on the fuselage at this time. After July 1, 1941, the "1-MB" part of the side number was obsolete, since VMB-1 was redesignated Marine Scout Bomber Squadron 132 (VMSB-132) at that time. *National Museum of Naval Aviation*

A formation of three SBD-1s from VMB-1 photographed on May 28, 1941, includes the aircraft of the commander of Section 3 in the foreground, aircraft number "7" of the squadron, which had a blue band around the fuselage. The next plane, number "4," was from Section 2, with a white band on the fuselage. The Bureau Numbers are, *foreground to background*, 1616, 1606, and 1619.
National Museum of Naval Aviation

The SBD-1 assigned to the squadron commander of VMB-2, BuNo 1626, is flying over a suburban area while based at Naval Air Station (NAS) North Island, San Diego, California, in 1940. The yellow paint of the wing top is discernible. The aircraft was painted overall in an aluminum-colored varnish.

The same Douglas SBD-1 seen in the preceding photo, BuNo 1526, is viewed from below during a flight in 1940. The plane bears the same markings as in the preceding photo, but the red cowling, fuselage band, and rudder stripe all appear to be almost white, due to the vagaries of the vintage black-and-white photo processing. *National Museum of Naval Aviation*

BuNo 1526 is seen from the left during a flight over Southern California. Note the pitot tube mounted under the left wing, and the external arrestor hook on the bottom of the fuselage forward of the tail landing gear.

At the time this photo of SBD-1 BuNo 1597 was taken at NAS Anacostia, Washington, DC, in August 1940, this plane was serving as the assigned aircraft of the commander of VMB-2. A telescopic sight, not present in the preceding photos of BuNo 1526, has been mounted in the windscreen of this plane. *National Museum of Naval Aviation*

Marine Bombing Squadron 2 was transferred from San Diego to the 2nd Marine Aircraft Group (MAG-2), Fleet Marine Force, based at Marine Corps Air Station Ewa, on Oahu, in January 1941, traveling to Hawaii aboard USS *Enterprise* (CV-6). In this photograph, SBD-1 BuNo 1736 from VMB-2 has just landed on the *Enterprise* during that transit. *National Museum of Naval Aviation*

Perforated flaps lowered, an SBD-1 from VMB-2 prepares to land on USS *Enterprise* during 1941. At the time the squadron was transferred to MCAS Ewa in January 1941, it was equipped with twenty SBD-1s. *National Museum of Naval Aviation*

In a vintage, undated color photo, SBD-1 number 4 from Marine Scout Bombing Squadron 132 (VMSB-132) flies a training mission over land from its base at Naval Air Station Quantico, Virginia. The plane has a black spinner and Light Gray camouflage paint, with discoloration in the paint surface from fading and wear. Note the radio direction-finder loop antenna between the front and rear seats. *Stan Piet collection*

A view of three SBD-1s from VMSB-132 includes in the foreground the same Dauntless featured in the preceding photo, number 4. The other two planes are numbered 14 and 11. Note the recessed steps on the sides of the fuselages adjacent to the front cockpits. Similar steps are adjacent to the rear cabins but are not so easily discerned.
Stan Piet collection

A photo dated April 20, 1943, shows a Douglas SBD-1 after crashing on a beach at or near Daytona, Florida. A partial side number, 0-S-03, is visible. This plane may have been assigned to the Daytona Beach Operational Training Unit. It is known that Ens. J. M. Wallace and his observer/gunner, of that unit, were killed on April 6, 1943, when their SBD-1, BuNo 1621, failed to pull out of a dive off Ormond Beach, Florida. *National Archives*

CHAPTER 2

SBD-2

The fuel supply of the SBD-1's successor, the SBD-2, was increased to 310 gallons, resulting in an increase in range from the 900 miles of the SBD-1 to 1,200 miles. The additional fuel was housed in a 65-gallon fuel tank in each outer wing panel of the Dauntless. At the same time, the two 15-gallon tanks previously found in the center section of the wing were deleted.

Armament of the Dauntless was unchanged from the previous model, consisting of up to 1,200 pounds of bombs, a flexible .30-caliber machine gun operated by the rear-seated gunner / radio operator, and a pair of forward-firing .50-caliber machine guns mounted in the upper cowl and synchronized to fire through the propeller arc. The latter were fired by the pilot.

The rear-seat crewman, in addition to manning the gun, also served as radio operator. Additionally, he had a set of rudimentary flight controls.

Even though the SBD-2 lacked armor, the Navy rostered the type, and in fact SBD-2s became some of first Navy aircraft to come under fire. On December 7, 1941, three SBD-2s assigned to USS *Lexington* (CV-2) were not aboard ship, instead sitting at a Pearl Harbor airfield when the enemy struck. While two of the three were destroyed, the third not only survived that attack but remarkably survives to this day. It is now on display at the National Museum of Naval Aviation at Pensacola, Florida, and photographs of that aircraft aid in illustrating this chapter.

Three days after the Pearl Harbor strike, SBD-2s from USS *Enterprise* CV-6 drew Japanese blood in retaliation, sinking Imperial Japanese Navy (IJN) submarine *I-70*. In that action, Ens. Perry L. Teaff, flying an SBD-2 of Scouting Squadron (VS) 6 from USS *Enterprise* (CV-6), sighted *I-70* on the surface just after 0600. Launching an attack, Teaff scored a near miss with a 1,000-pound bomb and inflicted enough damage so that *I-70* could not submerge.

That afternoon, another VS-6 Dauntless, this one flown by Lt. (j.g.) Clarence E. Dickinson Jr., who himself had been shot down and his gunner killed on December 7, again located the *I-70*. Dickinson climbed to 5,000 feet and launched a dive-bombing attack. Despite evasive action and antiaircraft fire from the sub, Dickinson was able to put his bomb right beside the *I-70*, amidships. The blast hurled several of the Japanese antiaircraft gunners overboard, and *I-70* began to settle immediately. Forty-five seconds after the attack, only bubbles of oil and foamy water, and four survivors, remained.

Total production of the SBD-2 was eighty-seven aircraft, BuNos 2102 through 2188 inclusive.

However, only eighty-six made it onto Navy rolls. BuNo 2109, the eighth SBD-2, crashed during acceptance trials. The Bureau Number was reassigned to a replacement aircraft, which was an SBD-3.

The second production model of the Dauntless, the SBD-2, introduced several improvements. Although this plane lacked crew armor and was still not considered ready for combat, it had a fuel capacity of 310 gallons: 100 gallons more than the SBD-1. This was achieved by eliminating two 15-gallon fuel tanks and installing two 65-gallon fuel tanks in the outer wing sections. Eighty-seven SBD-2s were produced, BuNos 2102 to 2188. The SBD-2 depicted in the following factory photos was the first one, BuNo 2102. *National Archives*

As seen in this frontal view, the principal visual difference between the SBD-1 and SBD-2 was the lower profile of the carburetor air scoop on the SBD-2. Projecting from the lower quarters of the cowling are the exhausts. *National Archives*

The perforated diving flaps are visible in this rear view. The hydraulically powered landing and diving flaps of the SBD-1, and other Dauntless models, included five perforated sections, built of ribs and stringers and clad with stressed skin on the outside only. Three of the sections, on the lower surfaces of the wing, acted as combination landing and diving flaps. Two more sections on the top rears of the inner wing panels deflected upward to serve as diving flaps in combination with the lower flaps. *National Archives*

The horizontal stabilizer and vertical fin incorporated ribs, stringers, and two spars, covered with stressed skin. The elevators and rudder had metal frames clad with fabric. Likewise, the ailerons were metal with a fabric covering. *National Archives*

SBD-2 BuNo 2012 is seen from the right side, wearing for markings only some small stencils and decals, its model designation on the rudder, its Bureau Number on the vertical fin, and "U.S. NAVY" under the horizontal stabilizer. On the side of the fuselage a few feet from the rear of the aft cockpit canopy is the door for the baggage compartment.

The cowl flaps (three per side) were hydraulically operated. Below them is the engine exhaust. Below the base of the antenna mast is an air-outlet vent in the closed position. Below the fuselage are the bomb crutch and its mounting post. *National Archives*

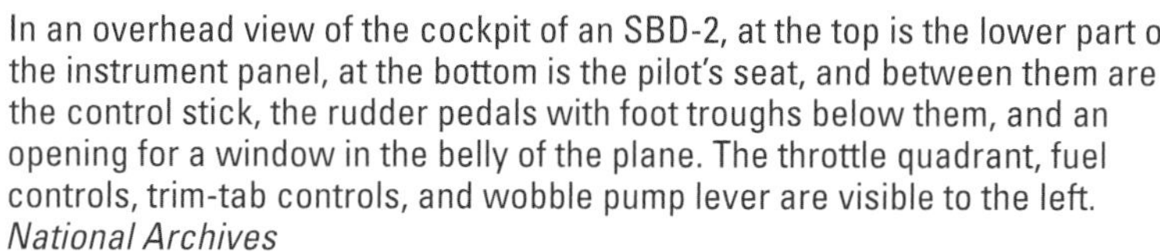
In an overhead view of the cockpit of an SBD-2, at the top is the lower part of the instrument panel, at the bottom is the pilot's seat, and between them are the control stick, the rudder pedals with foot troughs below them, and an opening for a window in the belly of the plane. The throttle quadrant, fuel controls, trim-tab controls, and wobble pump lever are visible to the left. *National Archives*

The instrument panel of the SBD-2 included a large upper panel, a smaller panel at the bottom, and a sliding chart table between them. The panel included standard items such as an altimeter, tachometer, manifold-pressure gauge, airspeed and turn-and-bank indicators, clock, and the control units for the directional gyro and the bank-and-climb unit. On the left side of the panel, the third instrument from the top was the checkoff instrument, a pilot's checklist readout display. The openings to the sides of the upper part of the instrument panel were for access to and clearance for the receivers of the two .50-caliber machine guns. *National Archives*

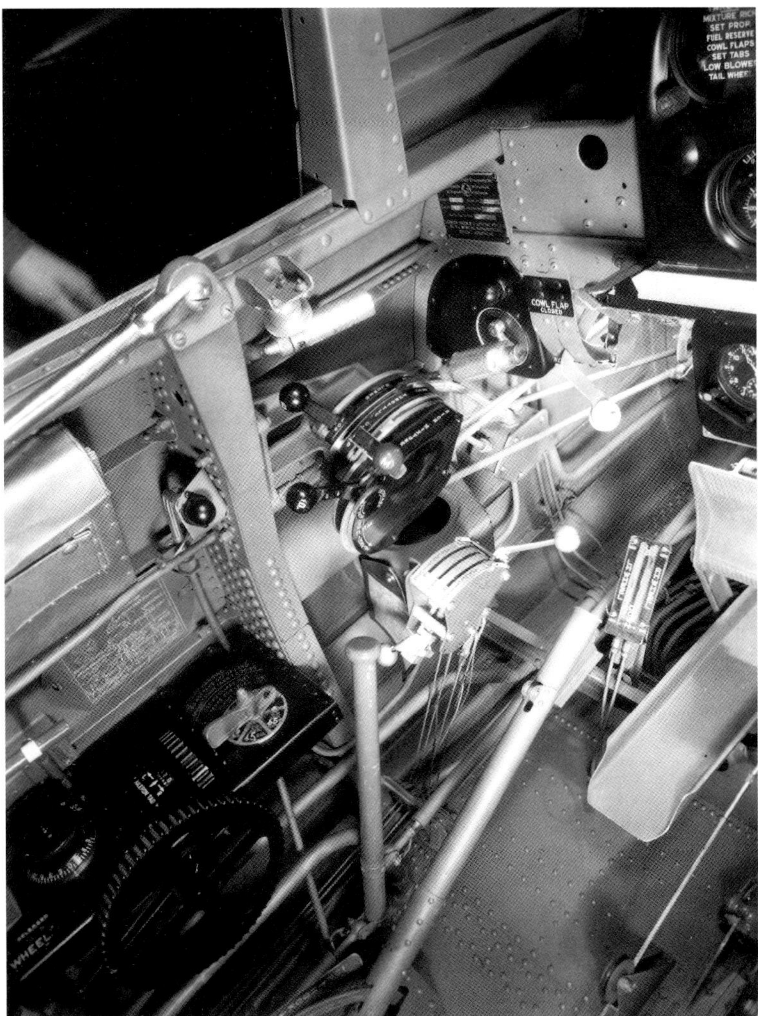

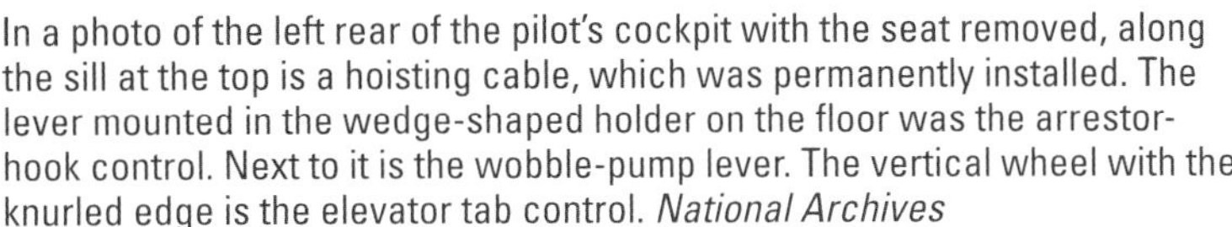

In a photo of the left rear of the pilot's cockpit with the seat removed, along the sill at the top is a hoisting cable, which was permanently installed. The lever mounted in the wedge-shaped holder on the floor was the arrestor-hook control. Next to it is the wobble-pump lever. The vertical wheel with the knurled edge is the elevator tab control. *National Archives*

Farther forward on the left side of the cockpit, with the seat removed, at the front of the console is the fuel tank selector. The throttle quadrant contains, *left to right*, the blower control, throttle control, and fuel-mixture control. Below the throttle quadrant is another quadrant, with the bomb-arming control and the bomb release control. To the right are the left rudder pedal and footrest. *National Archives*

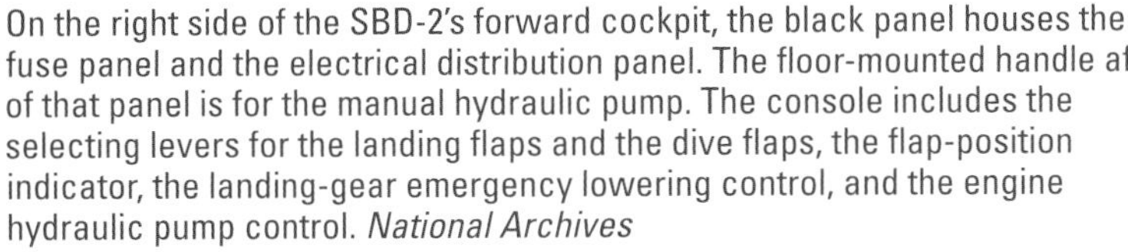
On the right side of the SBD-2's forward cockpit, the black panel houses the fuse panel and the electrical distribution panel. The floor-mounted handle aft of that panel is for the manual hydraulic pump. The console includes the selecting levers for the landing flaps and the dive flaps, the flap-position indicator, the landing-gear emergency lowering control, and the engine hydraulic pump control. *National Archives*

In a view of the right rear of the pilot's cockpit, the vertical tube is the right seat support. The T-shaped handle on the right side of the lower part of that support is the headrest release. Also in the rear corner of the cockpit are two racks for the pilot's oxygen rebreather. *National Archives*

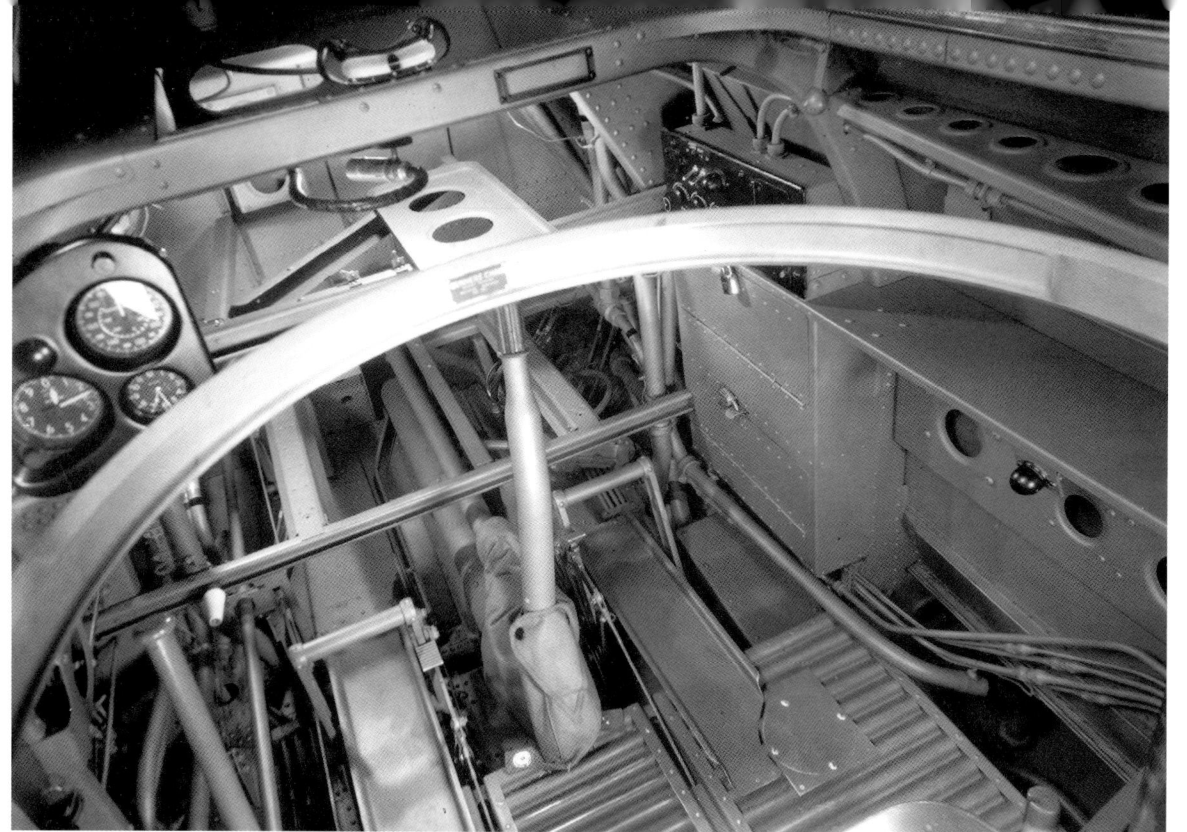

All models of the Douglas Dauntless included basic flight controls for the observer/gunner, which were useful to have in case the pilot were killed or disabled and the rear-seat man had to fly the plane. This view from the left rear of the aft cockpit shows the rear control stick, which was detachable, and the simplified, bar-type rudder pedals; to the left is a small instrument panel with an airspeed indicator, altimeter, and clock. *National Archives*

In a view more to the left front of the aft cockpit, to the rear of the small instrument panel is a throttle control, below which is a floor-mounted lever: the aft wobble-pump handle. *National Archives*

The aft cockpit is viewed from the left side, looking toward the right rear, with part of the observer/gunner's seat visible to the left. The carriage for the .30-caliber flexible machine gun is the object with the D-shaped locking handle located on the semicircular skate rail spaced out a few inches to the rear of the ring, which also supports the rear seat. To the rear of the ring are ammunition racks; below these racks are two footrests fabricated from metal rods. *National Archives*

The curved skate rail and other features of the aft cockpit are seen from the right side of the fuselage. Note the gun tunnel to the upper left, the footrests at the rear of the floor, and the first-aid kit on a rack toward the right of the photo. *National Archives*

Fourteen of the eighty-seven SBD-2s were converted to photoreconnaissance aircraft, designated SBD-2P. This example, assigned to VS-6 (based on USS *Enterprise*), is shown while conducting photographic mapping over the island of Molokai, Hawaii, on August 20, 1941. *National Museum of Naval Aviation*

This SBD-2 flying over the Pacific off Hawaii on October 10, 1941, is marked "COMMANDER AIR GROUP" in the shadow on the aft fuselage. This was a reference to the assigned aircraft of Lt. Cdr. Howard L. "Brigham" Young, commanding officer of the USS *Enterprise*'s air group. *Naval History and Heritage Command*

Three SBD-2 Dauntlesses flying abreast in formation exhibit the low-visibility camouflage scheme introduced by the Bureau of Aeronautics for ship-based aircraft on December 30, 1940: overall Nonspecular Light Gray. Application of the national insignia also had changed: small-sized ones now were on each side of the aft fuselage, and larger ones were on the top of the left wing and the bottom of the right one. No side numbers are visible.

Nine Douglas SBD-2s from Scouting Squadron 5 (VS-5) are flying in formation off a coastline. This squadron served on USS *Yorktown* (CV-5) and had been reequipped with SBD-3s by December 1941. *American Aviation Historical Society*

This photo of nine SBD-2s from VS-5 was very likely taken moments from the preceding photograph. To the lower right is the squadron commander's plane, side number 5-S-1. The side numbers range up to 5-S-16. *National Archives*

This famous composite photo shows nine SBD-2s from VS-6 are flying over the Pacific off Hawaii, with their ship, USS *Enterprise*, below, on Navy Day, October 27, 1941. These Dauntlesses exhibit a new camouflage scheme introduced on October 13, 1941, consisting of Nonspecular Blue-Gray on the upper parts of the plane and Nonspecular Light Gray on the lower surfaces. Also, the usual hyphens separating the elements of the side numbers have been omitted on these planes. Practice bomb dispensers are visible under the right wings. *National Museum of Naval Aviation*

Flight deck crewmen nicknamed airdales are pushing an SBD-2 backward so that the tailwheel will engage the channel of the outrigger protruding from the edge of the deck. The Dauntless had experienced an engine malfunction, and pushing it onto an outrigger was an expedient to clear the deck so flight operations could continue unimpeded. *National Museum of Naval Aviation*

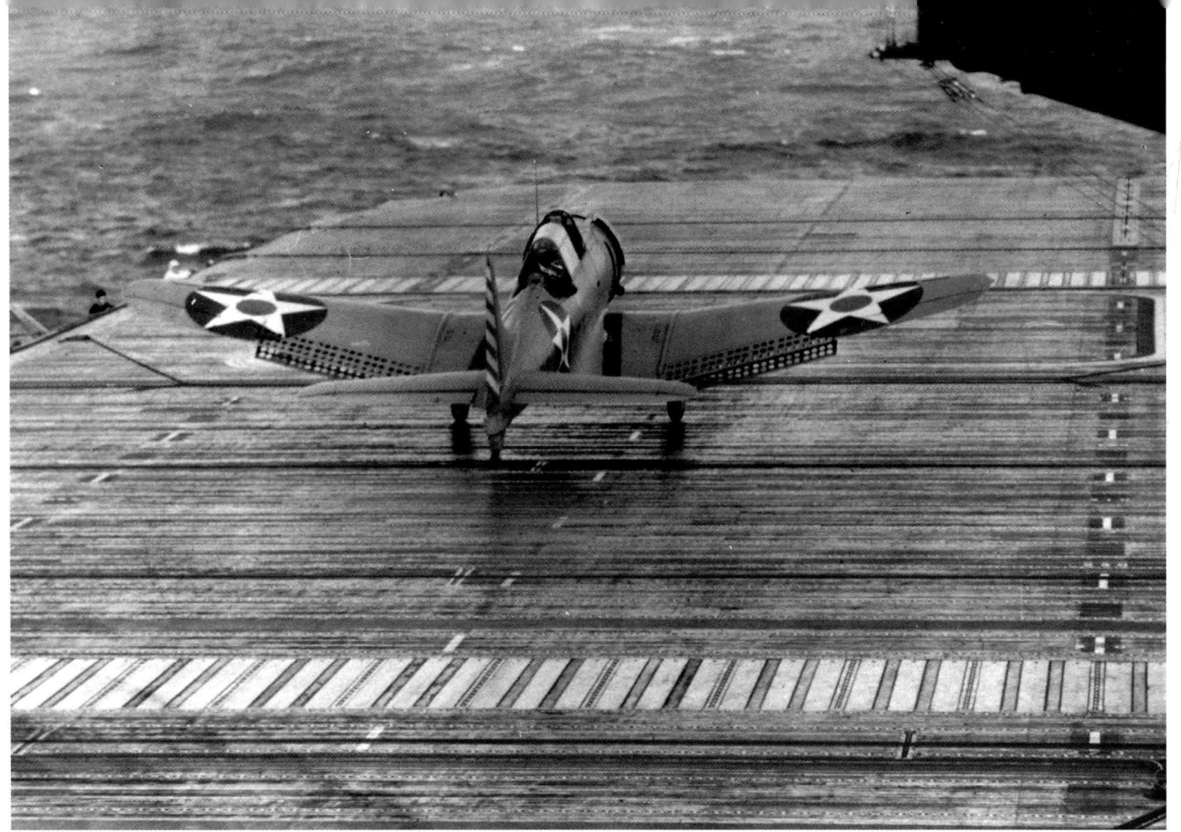

An SBD-2 from VB-6 is taking off from USS *Enterprise* at the start of the raid on Marcus Island on March 4, 1942. The very large national insignias on the wings and the fuselage were authorized in a directive from the Bureau of Aeronautics on December 26, 1941, as were the thirteen alternating red and white stripes on the rudder. *National Museum of Naval Aviation*

A fully bombed-up SBD-2 is poised for takeoff on the flight deck of USS *Enterprise* around May 1942. By this point, SBD-2s, never considered fully combat ready, were quickly being supplanted by the next generation of Dauntlesses, the SBD-3. Each plane has a large bomb on the centerline and a 100-pound on each pylon. *National Museum of Naval Aviation*

SBD-2s from VB-6 and VS-6 are being prepared for takeoff on USS *Enterprise* for a raid on Wake Island on February 24, 1942. These Dauntlesses have the Nonspecular Blue-Gray over Nonspecular Light Gray camouflage, and their side numbers omit the squadron number. The side numbers do have the "S" and "B" lettering for scout and bombing squadrons. *National Museum of Naval Aviation*

A mix of SBD-2s and SBD-3s are under construction at the Douglas Aircraft factory, El Segundo, California. Placards on the vertical fins identify the airframe by model, BuNo, and sequence number in the given model. The first four planes are SBD-2s, BuNos 2176 to 2179, sequence numbers 75 to 78. Next in line are three SBD-3s, sequence numbers 9 to 11, followed by more SBD-2s, starting with sequence number 79.

Douglas SBD-2 BuNo 2106 is a surviving Dauntless with an illustrious history. It was serving with VB-2 at Pearl Harbor on the morning of the December 7, 1941, attack and survived unharmed. Flying from USS *Lexington* (CV-2), the plane was part of the March 10, 1942, raid against Lae and Salamaua. Transferred to Marine Scout Bombing Squadron 241 in May 1942, the plane saw combat in the Battle of Midway, receiving over 200 bullet holes in the fight. While serving with a carrier-qualification training unit back in the States, on June 11, 1943, the aircraft was ditched and sunk in Lake Michigan. A salvage crew brought the plane to the surface in 1994, and it was restored to its current condition at the National Museum of Naval Aviation. Over 90 percent of the plane has original or period-correct parts. *National Museum of Naval Aviation*

Douglas SBD-2 BuNo 2106 is viewed from the left side with the perforated dive flaps partially elevated. On the wingtip in a small, streamlined fairing is the left, red navigation light. Mounted on the underside of the wingtip is the pitot tube. *National Museum of Naval Aviation*

Two .50-caliber machine gun muzzles protrude from the fuselage to the immediate rear of the cowling; to the fronts of the gun barrels are troughs in the top of the cowling, for clearance for bullets. The pilot's telescopic sight protrudes through the windscreen. *National Museum of Naval Aviation*

As seen through the left side of the front cockpit canopy, part of the instrument panel is in view, flanked by the butt ends of the receivers of the .50-caliber machine guns. An orangish-red eye cushion is affixed to the rear of the telescopic sight. Mounted under the top of the windscreen is a compass. *National Museum of Naval Aviation*

On the right side of the forward fuselage, just aft of the three cowl flaps, is a ventilation flap, in the open position. On the rear of this flap is stenciled, sideways, "PUSH HERE TO LATCH," with arrows pointing to the top and bottom rear corners of the flap. Also in view is the black nonslip walkway at the wing root.

The aft cockpit is viewed from the left rear, to the rear of which are the two doors for the gun tunnel, into which the flexible machine gun was lowered when not in use. To the rear of the doors is the blue lens of a lamp.

As built, SBD-2s had a single, flexible Browning .30-caliber machine gun in the aft cockpit. This shows a typical retrofit mount with twin .30-caliber machine guns of a type introduced during SBD-3 production. Under the sliding canopy sections is a loop-type radio direction-finding antenna.

The twin Browning .30-caliber machine guns are trained to the left, in this view taken from the left rear of the aft cabin. Above the rear of the barrel, the ring sight is folded down. To the immediate rear of the ring sight is a small armor plate, and another armor plate with an oblong sighting aperture is farther to the rear, above the gun's receiver

The twin Browning .30-caliber machine guns are trained to the left, in this view taken from the left rear of the aft cabin. Above the rear of the barrel, the ring sight is folded down. To the immediate rear of the ring sight is a small armor plate, and another armor plate with an oblong sighting aperture is farther to the rear, above the gun's receiver

Two rectangular recessed steps in the fuselage are provided on each side for the front and rear cockpits. Very faintly visible is the door to the baggage compartment on the rear half of the national insignia. A rounded fairing is fastened fore and aft over the joint between the outer wing section and the inner wing section. *Rich Kolasa*

The dive and landing flaps were actuated by torque tubes (visible in the preceding photo) with links (the white objects seen here) attached to the insides of the flaps. The torque tubes were powered by a hydraulic actuating cylinder housed inside the center wing section. When used as dive flaps, the lower flaps had a maximum depression of 42 degrees and the upper flaps had a 37-degree maximum elevation. *Rich Kolasa*

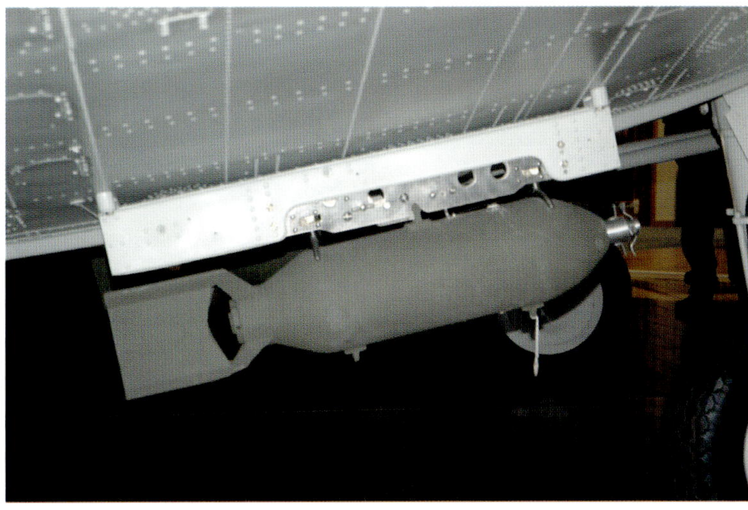

A 100-pound bomb is mounted on a pylon under the right wing. This type of pylon has a single shackle; the bomb has a single mounting lug on the topside for single-shackle pylons, and two lugs on the bottom for two-shackle pylons. Two sway bars attached to the front and the rear of the bare-metal part of the pylon keep the bomb from rocking from side to side while shackled to the pylon. *National Museum of Naval Aviation*

The center landing flap is viewed from the rear in the lowered position. The landing flaps had a maximum angle of depression of 45 degrees. The perforations in the landing and dive flaps were to prevent buffeting of the empennage when the flaps were extended.

The mounting post for the bomb crutch is viewed close-up from the right side. Above the post are two ejector ports; the large one is for spent .50-caliber cartridge cases and the small one is for ammunition links. *National Museum of Naval Aviation*

CHAPTER 3
SBD-3

Douglas Aircraft's SBD-3 was the first technically combat-ready model of the Dauntless, finally being equipped with crew armor, a bulletproof laminated-glass panel behind the windscreen, and self-sealing fuel tanks. The main exterior difference between the SBD-2 and SBD-3 was a change in the design of the ventilation flap on each side of the fuselage, aft of the cowling. That flap is in the open position in this photo of SBD-3 number 1, BuNo 4518, at Douglas Aircraft's El Segundo plant. Bureau Numbers for SBD-3s were 4518–4691, 03185–03384, and 06492–06701. *National Museum of Naval Aviation*

Although the SBD-2, as we have seen, did a commendable job despite its shortcomings, the SBD-3 was the first truly combat-capable version of the aircraft that was produced. Externally, the SBD-3 could be differentiated from its predecessor due to the presence of a revised, smaller ventilation slot just to the rear of the cowling. As production of the SBD-3 continued, and drawing from hard-earned combat lessons, the single rear .30-caliber machine gun gave way to a pair of the weapons. In time, earlier models of the Dauntless were retrofitted with the twin-gun arrangement.

With the pressure of war, production of the Dauntless was stepped up, and 584 of the SBD-3 variants were built. The Bureau Numbers assigned to the SBD-3 were 4518 through 4691, 03185 through 03384, and 06492 through 06701, along with the previously mentioned 2109, assigned the same BuNo as the destroyed SBD-2 that it replaced.

During the early war period, when fully half of the US Navy's carrier-borne combat aircraft were Dauntlesses, the SBD-3 bore the brunt of the offensive operations. Coral Sea, Midway, and the Eastern Solomons are but three of the actions in the Pacific that immortalized the Dauntless, and all fought with the SBD-3. On the other side of the globe, the SBD-3 took part in Operation Torch.

While some of the Operation Torch aircraft operating from USS *Ranger* swung four-blade propellers rather than the SBD-3's standard three-blade unit, the deck logs of the carrier make it clear—all the Dauntlesses aboard during that operation were SBD-3 models, as confirmed by their Bureau Numbers.

The fronts of the Hamilton Standard propeller blades of SBD-3 BuNo 4518 had three colored stripes on the tips, with the balance of the blades being the aluminum or silver color. Note the landing light on the bottom of the left wing, slightly inboard of the national insignia. *National Archives*

A left-rear view of the first SBD-3 includes details of the rudder and elevator trim tabs and their actuators. On the extreme rear of the fuselage below the rudder trim tab is a bulge: this was a fairing that was introduced with the SBD-3. *National Archives*

The small fairing on the right side of the tail is visible. The purpose of these fairings is not clear, and they evidently were not a full-production item, since they are not seen in subsequent photos of the SBD-3 and later models. Jutting at an angle from the tail, below the aforementioned fairing, is the attachment for a holdback cable, which, as the name implies, restrained the aircraft until the moment of release during a catapult launching. *National Archives*

SBD-3 BuNo 4518 is viewed from aft at Douglas Aircraft's El Segundo facility. From this angle, the difference in dihedrals of the center wing section and the outer wing sections is noticeable. *National Archives*

Like the SBD-2, the SBD-3 had a slightly elevated carburetor air intake on the top of the cowling. Note the slightly inward angle of the main-landing-gear oleo struts. *National Archives*

In a final view of the exterior of the aircraft, the model designation, SBD-3, is marked on the rudder, and the Bureau Number, 4518, is on the vertical tail. *National Archives*

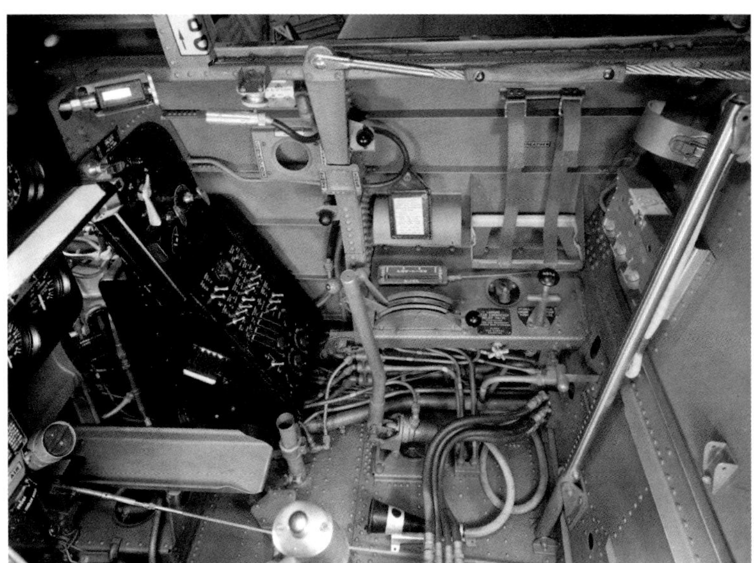

The right side of the pilot's cockpit of the SBD-3, shown with the seat removed, was similar to that of the SBD-1 and SBD-2, including the electrical distribution panel and fuse box, a small console, and the lever for the manual hydraulic pump. At the top center is the right anchor for the hoist cable. *National Archives*

Likewise, the left side of the pilot's cockpit remained virtually the same on the SBD-3. Below the hoist cable at the top is the map case. On the sidewall below the front of the map case is a diagram of the oil system. The two T-handles below and to the rear of that diagram are parachute-flare releases. *National Archives*

The pilot's instrument panel in the SBD-3 was unchanged from the SBD-2. In this example, the checkoff instrument is missing from its assigned opening on the left side of the panel.
National Archives

The Bureau Number on the tail of this Dauntless is indistinct but lies in the 03185–03384 range, indicating that this plane was a midproduction SBD-3. The photo was taken at El Segundo on March 4, 1942. The paint is Nonspecular Blue-Gray over Nonspecular Light Gray, with red and white stripes on the rudder. *National Museum of Naval Aviation*

The US Army Air Corps acquired a version of the Dauntless they designated the A-24 Banshee, based on the Navy's SBD-3. Some early examples the Army acquired carried the temporary USN designation SBD-3A, and the Army reportedly continued to use that designation for the duration of those planes' service. This SBD-3A in US Army markings, photographed at El Segundo on May 27, 1941, was marked "SBD-3A" on its tail. *National Museum of Naval Aviation*

The same SBD-3A is seen from the right rear. Faintly visible on the vertical fin is the Army serial number, AC41-15747, making it the second SBD-3A delivered to the Army. Army Banshees had large tailwheels and omitted the arrestor hook, and these features have been implemented on this SBD-3A. *National Museum of Naval Aviation*

In a photograph staged on the ground, an observer/gunner in full flight gear and oxygen mask is aiming the single .30-caliber machine gun in an SBD-3. During SBD-3 production the single .30-caliber machine gun in the aft cockpit would be replaced by a twin .30-caliber mount. *National Archives*

This photograph taken during the filming of the movie *G.I. Josie* shows the twin-.30-caliber Browning machine gun mount introduced during SBD-3 production. An ammunition feed chute is on the outboard side of each gun receiver. Three armor plates were fastened to the gun mount: two on the sides and one in the center with a vision slot for sighting the guns.

This twin .30-caliber machine gun installation was photographed in an SBD-3 at Douglas Aircraft's El Segundo plant on July 8, 1942. The guns were mounted on an Mk. 11 Adapter, a frame that included two grips with two triggers between them. Along with the twin machine gun mount came two sliding doors alongside the front of the gun tunnel, and the left sliding door is open in this photo. *National Museum of Naval Aviation*

An armorer is loading .30-caliber ammunition into a machine gun on an SBD-3 on the hangar deck of an aircraft carrier. A box in the rear of the aft cockpit held the .30-caliber ammo. The armor plates are not installed on this mount. *National Archives*

SBD-3 number "14" of Scouting Squadron 5, based on USS *Yorktown* (CV-5), bears the national insignia with the red circle in the middle omitted. This insignia was in use from May 6, 1942, to June 28, 1943. Note the practice-bomb dispenser under the wing and the white number "14" on the top of the wing. *National Museum of Naval Aviation*

Bombing Squadron 3 SBD-3s are running up their engines prior to taking off on a mission from USS *Saratoga* (CV-3) during the latter part of 1941. An F4F Wildcat from VF-3 is in the right foreground, while to the rear are several rows of Douglas TBD torpedo bombers, and then more Douglas SBDs. *National Museum of Naval Aviation*

On the flight deck of USS *Lexington* (CV-2), airdales are spotting an SBD-3 from VB-2 sometime during 1941. Note the ring-shaped holdback below the tail. *National Museum of Naval Aviation*

Crewmen are fueling an SBD-3 armed with an aerial depth charge on the flight deck of USS *Yorktown* on November 13, 1941. Standing on the deck is a man holding a fire extinguisher, in case a fire should suddenly break out. Two triangular brackets for attaching a gun camera are to the lower front of the windscreen. *National Archives*

An SBD-3 Dauntless serving with VS-2 speeds low over the water during a mission in 1942. Only the plane number, "10," and "S" for Scouting appear in the side number. The diagonal white stripe on the tail was called the LSO stripe and was intended to help the landing signal officer quickly assess if the aircraft was making a viable approach for a carrier landing. Below the windscreen is VS-2's Indian-head insignia. *National Museum of Naval Aviation*

With a Curtiss SO3C-1 Seamew in the lead, an SBD-3 of VS-5 flies a patrol along the coast of Oahu around mid-1942 to mid-1943. The SBD-3 has Yagi radar antennas under the wings, which were introduced to the production lines for the SBD-4 but were retrofitted on SBD-3s as well. *National Museum of Naval Aviation*

Douglas SBD-3s from VB-5 are being prepared for takeoff from USS *Yorktown* in April 1942. The first plane has its number in the squadron, "18," on the leading edge of each wing, above the main landing gear. Red and white stripes are visible on the rudder of that plane. *National Museum of Naval Aviation*

In contrast to the SBD-3s in the preceding photo, these Douglas Dauntlesses about to launch from USS *Yorktown* in April 1942 have their numbers marked on the fronts of their cowlings. In the first row is number "3," with number "9" to the right, and "6" in the left background. Note the absence of a propeller spinner on numbers "6" and "9." *National Museum of Naval Aviation*

An SBD-3 serving with VS-5 is being warmed up prior to takeoff from USS *Yorktown* on the first day of the Battle of Midway, June 4, 1942. This Dauntless was armed with the twin .30-caliber machine gun mount in the aft cockpit. Scouting Squadron 5 had been identical to Bombing Squadron 5 until redesignated VS-5 shortly before the Battle of Midway. *National Museum of Naval Aviation*

Plane number "15" from VB-3, based on USS *Yorktown*, has just made an emergency landing on USS *Enterprise* after being shot up during the first American attacks against the Japanese fleet in the Battle of Midway, on June 4, 1942. The right horizontal stabilizer and elevator have been severely damaged. *National Museum of Naval Aviation*

Airdales, most of whom are wearing steel helmets, suggesting they are in a combat zone, are spotting an SBD-3 from VB-8 on the flight deck of USS *Hornet* (CV-8), reportedly during the Battle of Midway in early June 1942. Running across the surface of the flight deck are metal tie-down strips, to which stays were attached for securing parked planes on the deck. *National Museum of Naval Aviation*

With the arrestor hook seemingly poised to catch a wire only a few feet in front of it, this SBD-3 from VS-41 is about to touch down on the flight deck of USS *Ranger* (CV-4) in August 1942. In the foreground on the deck is one of several supports for the arrestor wire, which held the wire above the deck to make it easier for the arrestor hook to catch it. *National Museum of Naval Aviation*

Members of the deck crew of the escort carrier USS *Santee* (CVE-29) are spotting Douglas SBD-3s from Escort Scouting Squadron 29 (VGS-29) on the flight deck in the Atlantic near the end of December 1942. Plane numbers are painted in black on the fuselages below the centers of the canopies, with the exception of number "11" in the background, on which the number is aft of the national insignia on the fuselage. All these planes have the sliding doors on the fuselage spines to the rear of the canopies, indicative of the presence of twin .30-caliber machine guns in the aft cockpits. *National Museum of Naval Aviation*

This SBD-3, number "12" of VS-41, has caught an arrestor wire, but with its nose-high attitude, it will be coming down on the flight deck with force. The scene was on USS *Ranger*, likely in or around August 1942. *National Museum of Naval Aviation*

Dive brakes raised, an SBD-3 marked as the twelfth plane of a scouting squadron is in a practice dive-bombing run. Mounted under the left wing is a practice-bomb dispenser, which carried several bomblets: a much-cheaper mode of practice bombing than full-sized bombs. *National Museum of Naval Aviation*

The propeller tips of these four SBD-3s spotted on the flight deck of USS *Ranger* in 1942 are painted black with red, yellow, and blue. The aircraft numbers are painted in white on each side of the fronts of the cowlings. The plane in the foreground has a Hamilton Standard Hydromatic propeller with a dome on the hub containing the pitch-changing mechanism. The other three planes have Hamilton Standard Constant-Speed propellers with flat hubs. *National Archives*

On the flight deck of a carrier, a crewman is test-firing twin .30-caliber machine guns. The national insignia has the yellow border associated with US aircraft participating in the November 1942 Operation Torch landings in North Africa. *National Archives*

A loop-type radio direction-finding (RDF) antenna is visible inside the aft cockpit of the nearest plane in a column of SBDs on a *Sangamon*-class escort carrier around the time of Operation Torch in November 1942. The old multicolored propeller tips have given way to yellow ones. *National Archives*

This SBD-3, BuNo 06624, with markings replicating those of the thirteenth plane of VS-41 when assigned to USS *Ranger*, is on static display at the Air Zoo, in Kalamazoo, Michigan. The plane crashed into Lake Michigan during a training flight on September 19, 1943, and was recovered and restored in the 1990s.

The left cowl flaps, engine exhaust, mast antenna, and bomb crutch are visible in this view of SBD-3 BuNo 06624. Note the air deflector in the open position on the rear of the fixed center section of the canopy. The deflectors, on both sides of the canopy, served to reduce the amount of airflow pouring into the aft cockpit when the two rear canopy sections were open.

The propeller spinner, parts of the propeller blades and hub, the front of the engine, and the carburetor air intake are viewed from the right side.

The bare-metal structure visible behind the right cowl flaps is the baffle ring, which is joined to the fuselage aft of the cowling.

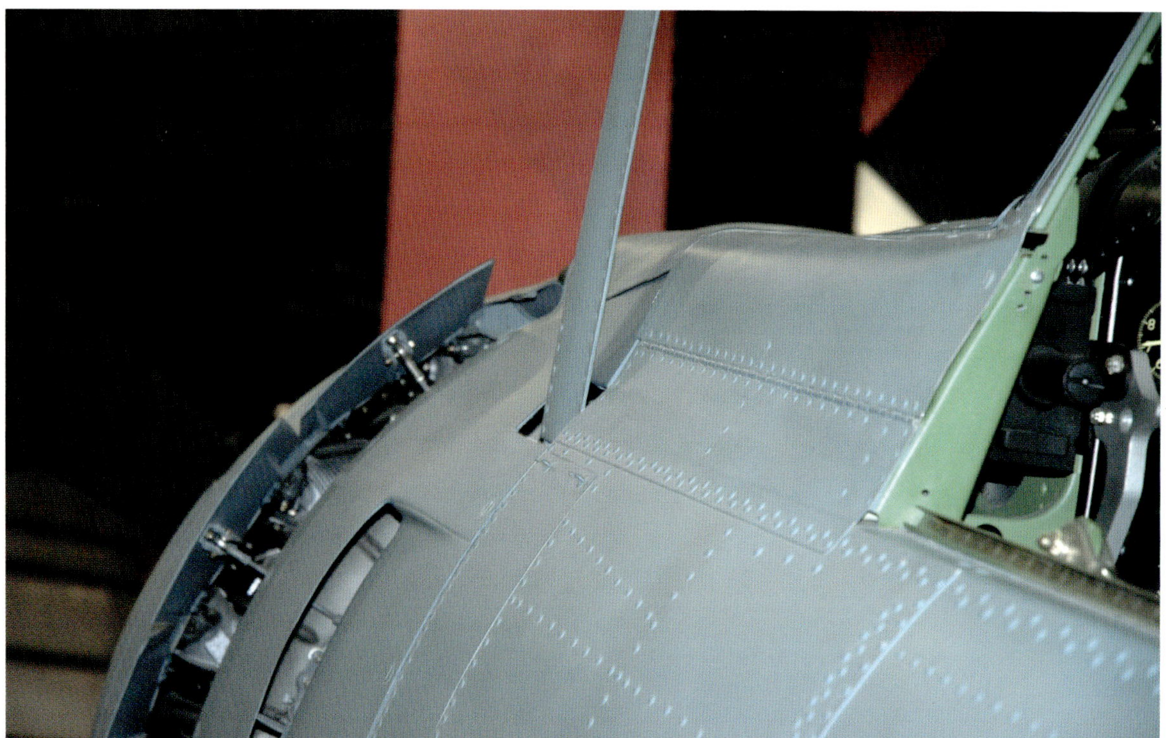

Two of the actuating links inside the left cowl flaps are in view, as is the open ventilating flap below the base of the mast antenna.

On the tip of the left wing is a navigation light with a red lens. Also in view are the base of the pitot tube, the letterbox slots in the wing, and the aileron hinges.

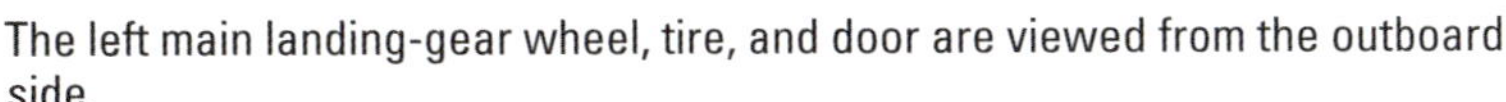

The left main landing-gear wheel, tire, and door are viewed from the outboard side.

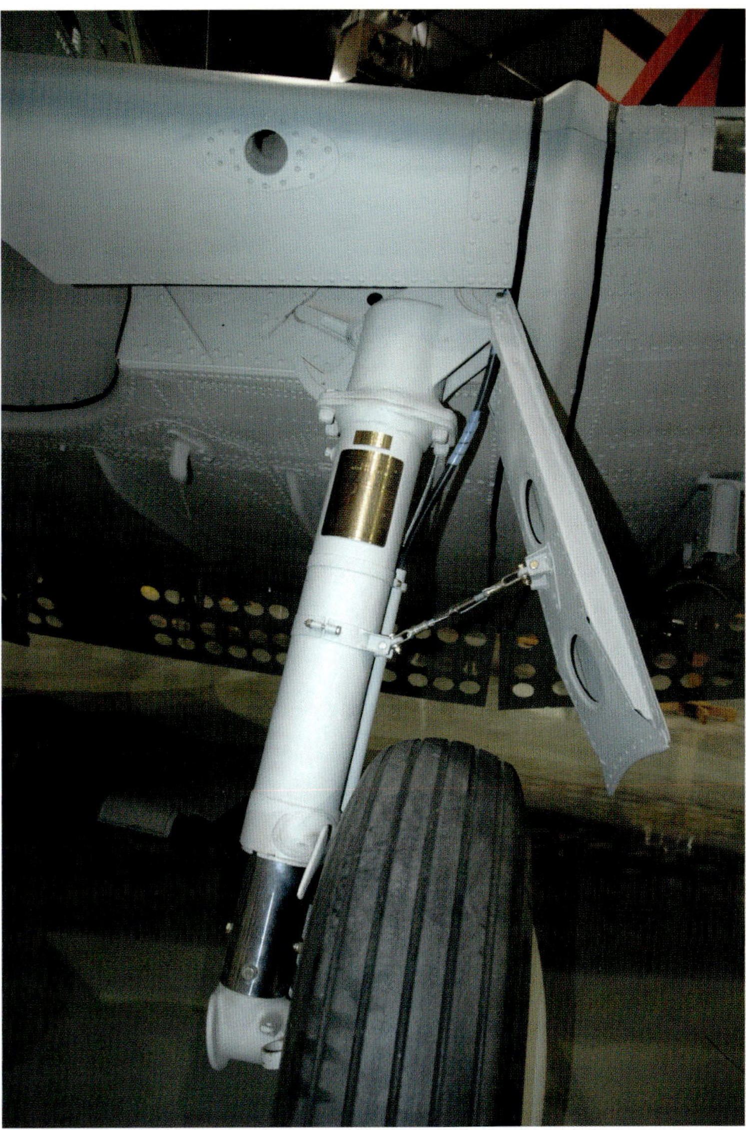

The left main gear of SBD-3 BuNo 06624 is viewed from the front. A drag link attached to the oleo strut served to move the door open or closed when the landing gear was lowered or retracted. Two brass data plates, a large one and a small one, are affixed to the oleo strut.

The left landing-gear bay is viewed from below and to the front. Inboard of the bay are two ejector chutes for spent .50-caliber cartridge cases and ammunition links. The major part of the bomb crutch is to the left. A fairing has been fastened over the recess in the belly of the plane, where the centerline bomb rack was located.

Above the rear of a bomb mounted on the centerline rack of the SBD-3 is the rear part of the recess in the belly of the plane, in which the Mk. 51 Mod. 7 bomb rack was installed. This rack could accommodate bombs weighing up to 1,600 pounds.

Details of the left rear of the bomb crutch are displayed. Above and to the rear of the end of the crutch is a door with a piano hinge, which was unscrewed and swung open when larger bombs were mounted on the centerline rack.

The oil-cooler air intake to the immediate rear of the bottom of the cowling was hinged to enable it to be opened, as seen here, or closed. To the front of the wing root are the ejector chutes for spent casings and links from the right fixed .50-caliber machine gun.

The dive flaps were attached to the wing with piano hinges, faintly visible here. The perimeters of the perforations in the flaps were tapered, with the holes wider at the top than at the bottom.

Details of the twin Browning .30-caliber machine gun mount and the fuselage and open canopy sections around the aft cockpit are displayed, including the recessed step below the number "41."

Elements of right side of the twin Browning .30-caliber machine gun mount are seen from the left side of the aft cockpit. To the lower left is the curved backrest of the observer/gunner's seat. Dangling from a cable next to the receiver of the right machine gun is a light-gray, T-shaped gun-charging handle.

The twin Browning .30-caliber machine gun mount is observed from the left side. The armor plates on the mount are painted an interior green similar to that seen in the interior of the aft cockpit. The Mk. 11 adapter and the ammunition feed chute are light gray. A clear view is available of the air deflector on the fixed part of the canopy. Below the canopy is an insulated lead-in for the wire antenna.

The ammunition feed on the left side of the twin .30-caliber machine gun mount is viewed close-up. The chute on the outer part of the feed assembly was designed to pivot, in order to keep the ammunition belt below the feed more or less straight as the guns were elevated or depressed. To the far left is the right-hand sliding door for the gun tunnel, a feature associated with the twin machine gun installations.

Seen from the left side of the lower part of the tail are the rear part of the arrestor hook, the tail landing gear, and the holdback fixture. Hidden in the tail are the struts and shock absorber for the tail gear.

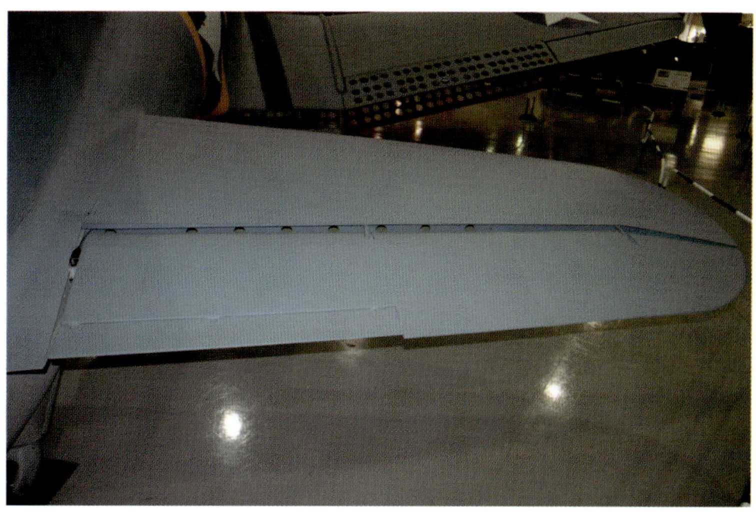

The right horizontal stabilizer, elevator, and trim tab are observed from the rear. Note the lightening holes in the rear surface of the stabilizer, partially visible above the front of the elevator.

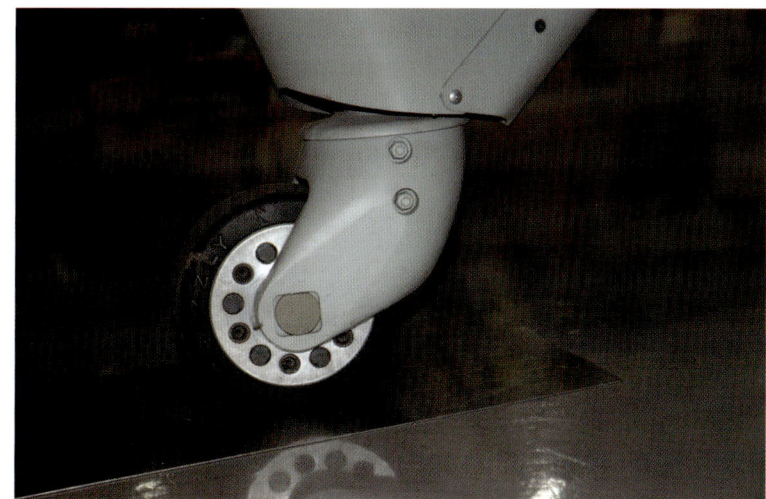

The solid-rubber tailwheel designed for aircraft carrier operations is mounted on the SBD-3. These wheels were 6 x 3 inches and, although not steerable, were fully castoring.

The arrestor hook, the tail landing gear, and the holdback fitting are viewed from the right side of SBD-3 BuNo 06624.

In regulation USN fashion, on the fabric-covered rudder is stenciled the plane's designation, SBD-3, while on the vertical fin are stenciled "NAVY" over the Bureau Number, 06624. The rudder trim tab is metal skinned.

The horizontal stabilizers of the SBDs were clad with aluminum alloy, as were the elevator trim tabs, while the elevators were covered with doped fabric.

CHAPTER 4
SBD-4

The fourth production model of the Dauntless, the SBD-4, differed in external appearance from its predecessor, the SBD-3, principally in its new propeller, the Hamilton Standard Hydromatic, with thicker blades with more-rounded instead of tapered tips than the earlier Hamilton Standard Constant-Speed propeller. The SBD-4 also had a new electrical system, upgraded from the earlier models' 12-volt system to 24 volt. has a small propeller and guard mounted on the side of the fuselage which powers the target tug winch. *Stan Piet collection*

In late 1942, the SBD-4 began to join the fleet. While externally almost indistinguishable from the SBD-3, the SBD-4 was actually much improved. Up front, a Hamilton Standard Hydromatic propeller replaced the old-type Hamilton Standard constant-speed model that had been installed at the factory on earlier models. The new propeller also had a wider chord, and the tips of the blades were more rounded. Less visible, the electrical system was changed from 12 volt to 24 volt, and electric fuel pumps were installed both for the primary and emergency systems. The catalyst for the adoption of a 24-volt electrical system was the desire to aid the installation of improved electronic gear, including directional aids as well as Yagi radar.

While most of the SBD-4s were assigned to the USMC, the Navy did place some in fleet service. The light aircraft carrier USS *Independence* (CVL-22) was notable in its use of the SBD-4 in late 1943. From her decks, the SBD-4 flew against Marcus, Wake, and Tarawa. Among the Marine units flying the SBD-4, VMSB-243, of the 1st Marine Air Wing, flew the type from Munda, New Georgia, in the Solomon Islands during August 1943.

Bureau Numbers were assigned to SBD-4s in two blocks: the first, BuNos 06702 through 06991, and later, 10371 through 10806.

Arrestor hook lowered and with a practice-bomb dispenser under the wing, an SBD-4 from Composite Squadron 22 (VC-22) approaches USS *Independence* (CVL-22) for a landing during the spring of 1943. The distinctive dome on the propeller hub, which housed the pitch-changing mechanism, was a key feature of the Hamilton Standard Hydromatic propeller. *National Museum of Naval Aviation*

SBD-4 number "10" from VB-42 cruises off a coastline during a training flight in the continental United States. Bombing Squadron 42 began accepting SBD-4s by early December 1942. *National Museum of Naval Aviation*

Airdales are spotting a Douglas SBD-4 assigned to VC-22 on the flight deck of USS *Independence* on April 30, 1943. The twin .30-caliber machine guns and their mount are not installed in the aft cockpit of this plane. *National Museum of Naval Aviation*

An SBD-4 from VB-41, side number 41-B-6, taxis on the flight deck of USS *Ranger* sometime in the first half of 1943. The plane's number, "6," is painted in white on the front of the cowling. *National Archives*

In Douglas Aircraft's El Segundo factory, a long row of SBD-4 Dauntlesses are under construction. Taped to the tails of most of the planes are signs with each plane's model designation and several identification numbers. Outer wing sections, horizontal stabilizers, elevators, and rudders remain to be mounted. *National Museum of Naval Aviation*

Douglas Aircraft workers are taking a coffee break alongside an assembly line of SBD-4 fuselages with the center wing sections installed. On most of the fuselages the baggage-compartment doors are hanging open. Projecting above the cockpits are the pilot's rollover structures. *National Museum of Naval Aviation*

Aboard USS *Independence*, airdales are pushing an SBD-4 backward on the flight deck on May 1, 1943. The plane was assigned to Composite Squadron 22. The outlines of the blades of the Hamilton Standard Hydromatic propellers are clearly visible. *National Museum of Naval Aviation*

Pilots working on their carrier qualifications are manning several SBD-4s on the flight deck of USS *Ranger* on June 20, 1943. Very large numbers have been painted roughly on the fuselages: number "11," in the foreground, has markings for VB-42, while numbers "4" and "2" in the background are marked for VB-41. *National Museum of Naval Aviation*

Assigned to VC-24, aboard the light carrier USS *Belleau Wood* (CVL-24), three SBD-4s, numbered "10," "12," and "13," fly an antisubmarine patrol around early July 1943. Aerial depth charges are on the centerline racks of all three Dauntlesses. The closest one has Yagi radar antennas under the wings and a bracket for a gun camera on the fuselage, to the front of the windscreen. *National Museum of Naval Aviation*

Five SBD-4s are proceeding on a practice-bombing mission; at least the closest one is equipped with practice-bomb dispensers. These planes are painted in the USN tricolor camouflage scheme of Sea Blue on upper surfaces, Intermediate Blue on the vertical tail and the middle areas of the fuselage, and Insignia White on the undersides. The national insignia with white side bars appears to have the red borders briefly used in the summer of 1943. *National Museum of Naval Aviation*

Spotted closely together on the flight deck of the light carrier USS *Cowpens* (CVL-25) during the ship's shakedown cruise in the Atlantic in July 1943 are seven SBD-4s and, with wings folded, three Grumman F6F Hellcats. The Dauntless in the left foreground is numbered "C5." *National Museum of Naval Aviation*

An SBD-4 from Marine Scout Bombing Squadron 233 evidently is being cannibalized for parts at Henderson Field, Guadalcanal. A number "4" has been painted roughly on the fuselage. Note the fuel tank inside the center wing section, one of two housed in that section. *National Museum of Naval Aviation*

CHAPTER 5
SBD-5

The SBD-5 was hands down the most numerous model of the Douglas Dauntless, with 2,965 units completed. Its main external difference from earlier models was the lack of a carburetor air intake on the top of the cowling. Also, the ventilation slot on each side of the fuselage was moved farther to the rear and was taller than on previous models. From this angle, the ventilation slot is almost in alignment with the mast antenna. *National Museum of Naval Aviation*

As American industry got fully on a war footing, Douglas established a new manufacturing facility in Tulsa, Oklahoma. One of the reasons this location was chosen was due to the tremendous demand on the Southern California labor pool. By shifting to midcontinent, not only was the labor situation eased, but the facility was also well beyond the range of enemy carrier-borne air strikes or bombardments, which was a concern at the time.

The first Dauntless model produced in the new plant was the SBD-5. The Tulsa plant would ultimately produce 2,409 SBD-5 aircraft. El Segundo produced a further 556 examples.

Among the several significant improvements introduced with the SBD-5 was the Wright R-1820-60 engine, with a rating of 1,200 horsepower, 200 more than the earlier model. Equally important, both the pilot and gunner were furnished with reflector sights. The importance of this change cannot be overstated. Earlier SBDs used a three-power telescope for bomb aiming. In the Pacific, the heat and humidity would cause these sights to fog, leading to the obvious problem of inaccuracy.

Visually, the three cowl flaps of the SBD-4 were replaced with a single cowl flap on the SBD-5.

The vertical air-outlet vent aft of the cowl on the SBD-5 was shifted to the rear and was notably taller than that of the SBD-4.

Sixty similar aircraft were ordered by the Navy for the US Army Air Forces, where they were intended to be designated A-24B-DE. Instead, the aircraft were delivered to the Marine Corps and subsequently designated SBD-5A instead. BuNos 09693 through 09752 were assigned to these aircraft. They could be distinguished from Navy aircraft by the lack of arrester hook, as well as by the use of a pneumatic tailwheel.

SBD-5 aircraft purchased for Navy use were assigned the following BuNos:

10807 through 11066

28059 through 29213

35922 through 36421

36433 through 36932

54050 through 54599

An SBD-5, BuNo 10957, marked "A.E.D.S." on the cowling and forward fuselage, is parked on a tarmac at an unidentified base on June 23, 1943. Another key feature introduced with the SBD-5 was the reduction in cowl flaps on each side from three to one; the right one is shown open here. *National Archives*

The same SBD-5 seen in the preceding photo is viewed from the left side. Mounted under the wing is a Yagi radar antenna. Equipment for this radar, including the scope, was installed in the rear cockpit. On the SBD-5, the pilot's telescopic gunsight was retired in favor of a more modern reflector gunsight. *Naval History and Heritage Command*

On October 1, 1943, SBD-5 BuNo 10892, piloted by Ens. J. H. Bushnell, dove off the flight deck of USS *Cabot* (CVL-28). The crew is believed to have been rescued. A large number "18" is on the fuselage, alongside the cockpits. *National Museum of Naval Aviation*

Flying above the battleship USS *Washington* (BB-56), with the fleet carrier USS *Lexington* (CV-16) in the background, an SBD-5 armed with an aerial depth charge on the centerline rack pursues an antisubmarine patrol. This was with Task Force 50 in preparation for a raid on the Gilbert Islands in November 1943. *National Museum of Naval Aviation*

Douglas SBD-5s from Marine Scouting Squadron 3 fly in formation during a patrol over the Caribbean in 1944. These Dauntlesses are painted in an Atlantic antisubmarine camouflage, with Nonspecular Dark Gull Gray on the top surfaces and alongside the cockpits, and Nonspecular Insignia White on the remainder of the aircraft. *National Archives*

Two airdales are securing an SBD-5 to tie-down strips on the flight deck of a carrier. The tops of the rope stays are secured to a retractable tie-down ring near the wingtip. Under the wing is painted in yellow the code R38. The national insignia on the fuselage has the red border authorized from July to September 1943. *National Archives*

A pair of petty officers are performing maintenance on the left main landing gear of the same SBD-5 seen in the preceding photo: note the code "R38" on the fuselage. The wheel cover has been removed, revealing more of the cast-magnesium wheel.

In a final color photo of the SBD-5 coded "R38," an airdale is attaching a grounding wire to the left exhaust. This was an important procedure to complete before a plane was refueled, to ensure that sparks from static-electricity buildup did not ignite fuel.

Douglas SBD-5s from VB-5, based on USS *Yorktown* (CV-10), are approaching Wake Island during an airstrike in early October 1943. The plane at the top has a large number "1" on the fuselage, and the national insignia that are in view have dark borders around them: the borders likely were painted blue after the red borders briefly in use in the summer of 1943 were deauthorized. *National Museum of Naval Aviation*

With Grumman Avengers flying above, four SBD-5s from VC-1 are flying an antisubmarine patrol over the Caribbean on October 3, 1943. These aircraft were based on USS *Cabot* (CVL-28), and the national insignia on the Dauntlesses have lighter-colored borders than the blue circles on the insignia, suggesting that the borders remained the Insignia Red color that the Navy had discontinued several months earlier. *National Museum of Naval Aviation*

The launch officer aboard USS *Yorktown* is an instant away from lowering his checkered flag, signaling the pilot of SBD-5 number "9" from VB-5 to take off. The photo was taken during the raids on Wake Island in early October 1943. These planes have a variation on the tricolor camouflage in which the darkest color, Sea Blue, extends down the sides of the cockpits. *National Museum of Naval Aviation*

Douglas SBD-5s have just returned to USS *Yorktown* after an October 1943 airstrike in the Pacific. Of particular interest is the unusual stencil on the tail to the right: "HANDS OFF." The plane in the background still has a bomb on the centerline rack. *National Archives*

Members of the crew of USS *Saratoga* are gathered around an SBD-5 numbered "53" from VB-12 in the aftermath of an airstrike on Rabaul in early November 1943. Stenciled in black on the right main landing-gear door is "NO PUSH." *National Museum of Naval Aviation*

Douglas SBD-5 number "45," with VB-16, lies with nose crumpled, propeller bent, right wingtip damaged, and left main landing gear collapsed, after crashing into barrier wires on USS *Lexington* in November 1943. The barrier, made of steel cables, was designed to stop a runaway aircraft that had failed to catch an arrestor cable during a landing. *National Museum of Naval Aviation*

The observer/gunner in a Dauntless snapped this photograph of an SBD-5, numbered "163" on the tail, on a mission near Munda, New Georgia, on January 18, 1944. The national insignia has the light-colored border associated with either the red border (discontinued five months earlier) or a red border repainted with a different shade of blue than that on the circular part of the insignia. *National Museum of Naval Aviation*

In March 1944 an SBD-5 marked "S-11" on the fuselage is flying an ASW patrol over the Caribbean. This Dauntless was operating from a land base with Scouting Squadron 37, which explains the large, pneumatic tire on the tail landing gear. *National Museum of Naval Aviation*

Task Force 58 lies below this SBD-5, bearing the number "2," from VMSB-231 at Majuro Atoll in the Marshall Islands in 1944. The pilot in this photo reportedly was Maj. Elmer Glidden, USMC. Below and to the rear of the base of the antenna mast is VMSB-231's insignia, featuring an ace of spades. Twenty-three bombs, symbolizing combat missions, are marked on the side of the front cockpit. *National Museum of Naval Aviation*

At an airfield on Majuro, in the Marshall Islands, SBD-5s are taxiing prior to takeoff on an airstrike against Japanese forces on Mili Atoll. First in line is a plane numbered B-1, with two 100-pound bombs under the wings and a 500-pound bomb on the centerline rack. For some reason, a dark swatch of paint has been applied to the lower half of the vertical fin and rudder of this plane. *National Museum of Naval Aviation*

A pair of SBD-5s with arrestor hooks in the down position are approaching USS *Enterprise*, part of Task Force 58, around early 1944. The Dauntlesses are numbered "6" and "18," and both of them have Yagi radar antennas under the wings. Note the down-hanging positions of the tail landing gears when there is no weight on them. *National Museum of Naval Aviation*

The Royal New Zealand Air Force was supplied with a quantity of SBD-5 Dauntlesses during World War II, including one with the serial number NZ5034 on the fuselage and the aircraft number, "34," on the tail, seen here being serviced on Espiritu Santo during March 1944. Although the plane was land based, the tailwheel retains the small, solid-rubber tire intended for carrier-based Dauntlesses. *National Museum of Naval Aviation*

Based on USS *Lexington*, SBD-5 number "37" from VB-16 flies near Dublon and Eten Islands on the way to a raid on the important Japanese base at Truk, on January 17 or 18, 1944. Five small bombs indicating combat missions had been drawn below the side of the windscreen. *National Museum of Naval Aviation*

Three SBD-5s from VS-31, including numbers "109" and "108," fly above the Pacific during a patrol mission off Samoa in the latter part of World War II. Note the unusual tricolor camouflage, in which the Sea Blue upper color curves down to the bottoms of the left bars of the national insignia, at least on the first two SBD-5s. All three planes have Yagi radar antennas under the wings. *National Archives*

A Royal New Zealand Air Force (RNZAF) Douglas SBD-5 Dauntless nicknamed "Winnie Puh III," serial number NZ5049, pursues a mission over the Pacific, carrying a full complement of bombs. The roundel on the bottom of the right wing seems to have been painted over with a dark color. The plane still has its arrestor hook, but a large, pneumatic tire has been mounted on the tail gear, in keeping with the RNZAF's land-based operations.

In late 1944, the United States sent a consignment of SBD-5s to the French, for use in the *Aéronautique Navale* (naval aviation). *Flotille* 4FB operated these planes during the final months of World War II. Following the end of the war, these Dauntlesses were assigned to the aircraft carriers *Dixmude* and *Arromanches*. Here, one SBD-5 of *Aéronautique Navale* is on the flight deck of a carrier, while another one is flying in the background.

The British were a very small-scale user of the SBD-5, the Royal Navy having acquired nine examples for evaluation in late 1943 or early 1944, designating them Dauntless I. These planes did not see operational service. Seen here is Dauntless I serial number JS997, painted in USN tricolor camouflage, with a US 58-gallon auxiliary fuel tank mounted on the left pylon.

Aéronautique Navale SBD-5 number "14" (former USN BuNo 54578) was in a heavily weathered and patched condition when photographed among the ruins of an airbase at Cognac, France, around the end of World War II. The French continued to operate Dauntlesses until July 1949.

Fuselages for SBD-5s are under construction on the assembly line at Douglas Aircraft's El Segundo, California, plant in August 1943. The photo provides an excellent view of the pilot's back armor, with mounts for the seat supports and the headrest. Note how the pilot's rollover structure still remained, behind the armor plate. To the sides of the dash in the front cockpit are the mounts for the .50-caliber machine guns; below them are the access doors for the .50-caliber ammunition compartments. *National Archives*

Another August 1943 photo at the El Segundo factory shows workers assembling SBD-5 fuselages in the foreground, while others are building center wing sections on jigs in the background. Those wing sections are oriented with their bottoms toward the camera, with the trailing edges of the wings to the top. *National Archives*

Douglas workers are handling a completed center wing section in August 1943. Visible inside the left side of the section is a fuel tank and plumbing. Directly below the tank on the bottom of the wing section is the left pylon, already installed. *National Archives*

In this area of the Douglas plant at El Segundo in August 1943, employees are assembling Wright R-1820 engines and their mounts, for installation in SBDs. In the foreground, workers are installing wiring on the accessories on the rears of the engines. These power plants are complete with their mounting frames and the bare-metal baffle rings. In the background are completed engine-and-mount assemblies, ready for installation on SBD fuselages. *National Archives*

Almost as far as the eye can see, fuselages for Douglas SBD-5s are lined up in the El Segundo factory in August 1943, awaiting the addition of engines, wings, and other components. In very small letters on the signs taped to the vertical tails are the Bureau Numbers for these fuselages: the first visible sign is for BuNo 28495, and the numbers go down sequentially from there to the background. *National Archives*

This surviving late-production SBD-5, BuNo 54532, which currently flies with the markings of the Commemorative Air Force, spent the decades after World War II among a number of private owners before being restored in the last half of the 1990s. It is shown here bombed up and prepared for a sortie at an airshow. *Rich Kolasa*

Douglas SBD-5 BuNo 54532 is painted to replicate the USN tricolor camouflage scheme and is decorated with the national insignia correct for the summer of 1943, with the red border around it. The plane's civil registration number, NL82GA, is visible below the horizontal stabilizer. *Rich Kolasa*

A view of SBD-5 BuNo 545342 from below emphasizes features such as the retracted main landing gear, the perforated landing flaps, the letterbox slots in the outer wings, the landing light on the left wing, and the numerous access panels on the bottoms of the wings. *Rich Kolasa*

Some three-quarters of a century since it first took to the air, SBD-5 BuNo 28536 still awes crowds with its aerial feats. This plane was delivered to the Navy in June 1943 and was shipped to the Royal New Zealand Air Force under the Lend-Lease Program. After flying in some thirty combat missions for the RNZAF under serial number NZ5062, this Dauntless was transferred to the US Marine Corps in May 1944. After the war, the plane, minus its original wings, served as a wind generator for MGM Studios in Hollywood, California. It was restored to flying condition in 1982. *Rich Kolasa*

The wings on SBD-5 BuNo 28536 are not original to this airframe but were recovered from Guadalcanal. The plane currently is in the collections of the Planes of Fame Air Museum. *Rich Kolasa*

Perforated landing flaps lowered, SBD-5 BuNo 545342 comes in for a landing. The vertical fin is supplied with an LSO stripe, a diagonal white marking designed to help landing signal officers visually determine if an approaching plane was in the correct attitude for a carrier landing. *Rich Kolasa*

SBD-5 BuNo 28536 warms its engine on an airport tarmac. The plane's current civil registration number, NX670AM, may be seen below the horizontal stabilizer. This plane has been equipped with a large, pneumatic rear tire, for land-based operations. *Rich Kolasa*

CHAPTER 6

Army Dauntless

With Germany's famed dive-bomber, the Junkers Ju 87 *Stuka*, grabbing headlines worldwide, it is little surprise that the US Army desired to add dive-bombers to their arsenal. To do so, the Army turned to the already developed Dauntless, borrowing some SBD-1 aircraft from the Marine Corps.

After a favorable evaluation by the 24th Bombardment Squadron, the War Department ordered seventy-eight similar aircraft, designating them A-24-DE, on September 27, 1940. In view of the fact that Navy contracts dominated production in the El Segundo plant, the most expeditious way to handle this transaction was to have the Navy order the aircraft on the Army's behalf. The Army assigned serial numbers 41-15746 through 41-15823.

The A-24, dubbed Banshee, differed little from its contemporary the SBD-3 but featured Army instrumentation and radio equipment. Further, it was equipped with a pneumatic tailwheel rather than the solid-rubber tire used by the naval version, and it notably lacked a tailhook.

Deliveries of the Army aircraft began on June 17, 1941. The 27th Bombardment Group was the first unit to be equipped with the A-24. The unit was being deployed to the Philippines when the Japanese struck Pearl Harbor.

As the Navy aircraft progressed to the SBD-4, the Army followed suit, the new version being designated A-24A. The A-24A retained the Army-specific components of the A-24, while incorporating the improvements in propellers and electrical system that came with the SBD-4. The aircraft, which were built in El Segundo and delivered from October 1942 to March 1943, were assigned serial numbers 42-6772 to 42-6831 and 42-60772 to 42-60881.

The final version of the A-24 was the A-24B-DT, which was based on the Navy SBD-5. Unlike the previous Banshees, which were ordered by the Navy at the behest of the Army, the A-24B-DTs were ordered directly by the US Army Air Forces (USAAF). Also differing from prior practice, these aircraft were built alongside B-24 Liberators in Douglas's plant in Tulsa, Oklahoma. Although 1,200 had been ordered in November 1942, 585 were ultimately canceled as the Army moved away from the dive-bomber. The 615 that were built were assigned Army serial numbers 42-54285 to 42-54899.

Some of the A-24Bs were assigned to 11th Air Force's 407th Bomb Group. These aircraft were used in attacks on the Japanese-occupied island of Kiska, Alaska, during July–August 1943.

Overall, the Army's experience with the aircraft did not parallel that of the Navy, and the type was soon relegated to training and utility use.

Based on the SBD-3, the Douglas A-24 Banshee was a US Army dive-bomber that did not see extensive production or much combat, since the Army did not stress dive-bombing as a major tactical concept. Because the A-24 was designed for land-based operations, it lacked an arrestor hook and had an inflatable rubber tire on the tail landing gear. Here, an A-24, *left*, painted in Army camouflage of Olive Drab over Neutral Gray, is parked next to an SBD-3 at Douglas Aircraft's El Segundo factory on June 20, 1941. *National Museum of Naval Aviation*

"ARMY" is marked under the left wing of an A-24 at an unidentified site. Also visible under that wing are a bomb rack, the landing light, the three letterbox slots, and the pitot tube. *National Museum of the United States Air Force*

Although the arrestor hook was omitted from the Douglas A-24, the mounting bracket for it remained in place on the bottom of the fuselage. Unfortunately, the tail number on this aircraft is too small to be read. *National Museum of the United States Air Force*

As seen in a left-rear view of the same A-24, white decals marked "STEP" are affixed above the recessed steps alongside the two cockpits. On the round access door farther aft on the fuselage is a decal or sticker. *National Museum of the United States Air Force*

From the front, the A-24 is virtually indistinguishable from the SBD-3, including the style of the carburetor intake on the upper front of the cowling. *National Museum of the United States Air Force*

An A-24 Banshee is observed from the left rear with the cockpit canopies open. The small, solid-rubber rear wheel of the SBD-3 would not have been very effective while taxiing on a grassy field such as this one. *National Archives*

Three Douglas A-24s fly in formation with only the pilots present. The large, inflatable tires of the A-24 tail gear are particularly prominent in this photograph. *National Archives*

In this view of a US Army A-24 Banshee in a dive, the dive flaps cast a large shadow on the forward fuselage. This photo apparently depicts a training or test flight, since there is no observer/gunner present in the rear seat. *National Museum of the United States Air Force*

Three A-24 Banshees are flying in formation, probably during a shakedown flight over Southern California sometime between mid-1941 and mid-1942. The Olive Drab over Neutral Gray camouflage scheme is seen to good effect in this color photograph. *Stan Piet collection*

Newly completed Douglas A-24 Banshees are lined up inside the Douglas Tulsa, Oklahoma plant, with a B-24 parked in the background. The A-24s were equipped with the Hamilton Standard Constant-Speed propellers and with spinners. The national insignia are of the design used from May 1942 to June 1943: white star, blue circle, and no red circle in the middle. *National Museum of the United States Air Force*

The Douglass A-24A Banshee was similar to the A-24, except it was based on the SBD-4. As a result, the A-24A was equipped with the Hamilton Standard Hydromatic propeller, as seen here. Most of the other differences between the A-24 and the A-24A were not noticeable from the outside. Shown here is an A-24A at Selman Field, Louisiana, before February 1944. The serial number, 42-6795, is marked on the vertical fin. *National Archives*

Based on the SBD-5, the Douglas A-24B Banshee lacked the carburetor air intake on the top front of the cowling and had a single cowl flap on each side and a repositioned and redesigned air vent on each side of the fuselage aft of the cowling. This bare-aluminum example, A-24B serial number 42-54736, photographed at Morotai, Dutch East Indies, on January 1, 1945, was equipped with a radio direction-finding (RDF) "football" antenna under the fuselage. *National Museum of the United States Air Force*

Led by a guide Jeep with a sign reading "FOLLOW ME" on its rear, an A-24B, serial number 42-54459, is taxiing to a dispersal area on Makin in the Gilbert Islands on December 13, 1943. The man standing on the wing next to the pilot's cockpit was likely the plane's observer/gunner. *National Museum of the United States Air Force*

Whereas the French Navy operated SBD-5s, the French *Armée de l'Air* (air force) flew A-24Bs, one of which is depicted here. Supplied with fifty of these planes, the French distributed half of them to Meknes airbase in Morocco and the balance to two squadrons: GB 1/17 "*Picarde*," and GB 1/18 "*Vendée*." *National Museum of the United States Air Force*

The nicknamed "Dry Run" is emblazoned on the nose of this A-24B from an unidentified unit at a forward airfield: note the matting on the runway. A 100-pound bomb is shackled to each wing pylon, and several larger bombs without the fins installed are lying on the ground next to the right main landing gear. Sandbags propped against the wheels serve as chocks. *National Museum of the United States Air Force*

Fuerza Aérea Mexicana (Mexican air force) operated A-24Bs during and after World War II. This example exhibits postwar characteristics, including a bare-aluminum finish, a blue stripe along the fuselage, and tinted glass on the upper part of the canopy. Under the wing is the triangular, red-white-green national insignia. The tail number is BID-2525. *National Museum of the United States Air Force*

The Pima Air & Space Museum, Tucson, Arizona, preserves this Douglas A-24B. This model of the Banshee was based on the SBD-5 and thus lacked the carburetor air intake on the top of the cowling.

Douglas A-24B, serial number 42-54582, is displayed, complete with mannequin crewmen, at the National Museum of the US Air Force, Dayton, Ohio. An excellent view is provided of the open dive flaps and landing flaps.

The open canopy sections, the twin Browning .30-caliber machine guns, and the round door for the storage compartment for an inflatable life raft and emergency rations of A-24B, serial number 42-54582, are observed. The twin .30-caliber machine guns in the A-24B at the National Museum of the US Air Force have the early-style cooling jackets with elongated cooling holes instead of the round holes appropriate for this model of Banshee. The ammo boxes also were not used on the twin machine gun mounts. The small sliding door on the deck to the rear of the aft cockpit, however, is correct.

The letterbox slots and the base of the pitot tube are viewed close-up. Note the rolled edges at the rears of the slots.

A small bomb is mounted on the left pylon of an A-24B, serial number 42-54582. Also in view are the numerous access panels on the bottom of the wing, and the main landing gear.

Viewed from the left front of the A-24B are the air scoop for the oil cooler in the open position, a bomb on the centerline rack, the left wheel well, and the ejector chutes for the left .50-caliber machine gun.

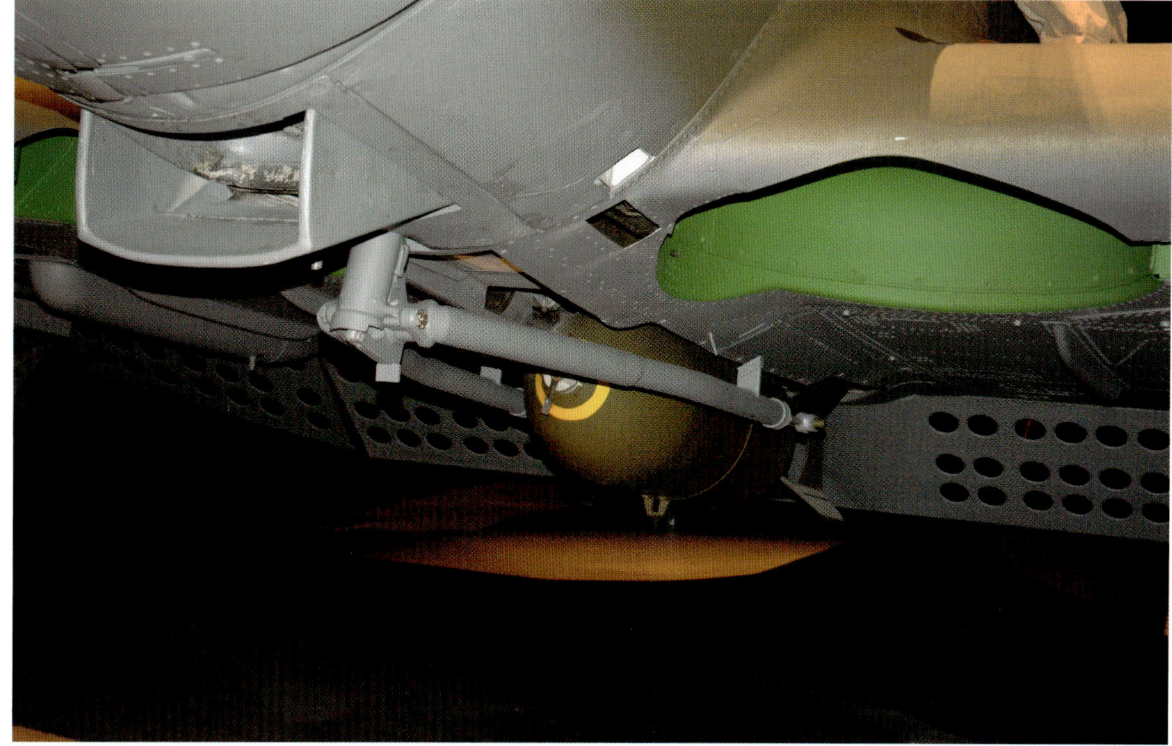

In a view of the left side of the empennage of the A-24B Banshee at the National Museum of the US Air Force, the bracket for the arrestor hook, which was not used on the Banshees, is present on the bottom of the fuselage.

CHAPTER 7

SBD-6

The final production model of the Douglas Dauntless was the SBD-6. Visually, it was almost identical to the SBD-5, with the lack of a carburetor air intake on the top of the cowling and the revised air vent on each side of the forward fuselage, in line with the antenna mast. The SBD-6 was distinguished from the SBD-5 by the change from the Wright R-1820-60 engine to the R-1820-66, which yielded an increase in horsepower from 1,200 to 1,350. *American Aviation Historical Society*

The final production variant of the Dauntless was the SBD-6, which was the most powerful version of the type, boasting 1,350 horsepower being delivered by a Wright R-1820-66 Cyclone. The test article of this version, the XSBD-6, was converted from SBD-5 BuNo 28830.

Work on the proposed replacement for the Dauntless, the Curtiss SB2C Helldiver, had begun in 1939. In the spring of 1942, after only a single prototype of the SB2C had flown (and crashed), the Navy announced that 3,000 Helldivers had been ordered from Curtiss. This was augmented by orders for a further 1,000 from Canadian Car and Foundry at Fort William, Ontario. However, it would be May 1943, well beyond the originally planned service date, before the Helldiver was ready for fleet trials. Unfortunately, those were less than 100 percent successful, with famed aviator and USS *Yorktown* (CV-10) skipper "Jocko" Clark flatly refusing to take the new Helldivers into combat, demanding that the venerable Dauntless be placed back aboard.

By the time production of the SBD-6 began in early 1944, the problems of the SB2C had been ironed out. Accordingly, only 450 of the SBD-6s were built before Dauntless construction ceased.

The same SBD-6 is viewed from the right rear. The Bureau Number on the vertical fin is 54689, which coincided with the ninetieth SBD-6 completed. SBD-6 production was cut off after 450 examples, and most, but not all, of them served as training aircraft in the United States. *National Archives*

Power plant assemblies on trolleys are ready for installation in SBD-6s at Douglas Aircraft's El Segundo plant on July 10, 1943. The nearest example, viewed from the right rear, shows how each assembly included, from the rear forward, the engine mount (with the oil tank and the oil cooler between the legs of the mount), the baffle, the exhausts, the sole cowl flap on each side, and the Wright R-1820-66 radial engine. *National Archives*

A March 31, 1944, photograph shows a Douglas SBD-6 from the right front. The pylons on this plane were painted the same Insignia White as the bottom of the wings, whereas in the past the pylons often had exhibited a bare-metal finish. Note how the auxiliary fuel tanks have received a two-color camouflage: likely Sea Blue over Insignia White. *National Museum of Naval Aviation*

In the Douglas El Segundo plant, a newly minted SBD-6 appears to be ready for delivery. The usual Douglas Aircraft identification placard is resting on the propeller. Striped, triangular warning signs are attached to the Yagi antennas and are marked "STAY CLEAR." The number "2000" is marked in white on the forward fuselage and the vertical tail. *National Museum of Naval Aviation*

The fifth SBD-6 completed, BuNo 54605, is seen from the right side. A bomb is mounted on the centerline rack, and the aircraft number, "15," is painted in white on the front of the cowling and the top of the tail. Tricolor camouflage has been applied to the aircraft.

Douglas SBD-6 airframes of Marine Air Group 24 have been consigned to a salvage yard at Balaban, Mindanao, in the Philippines on July 24, 1945, prior to scrapping. The engines have been removed. The nearest Dauntless has 114 bombs on its combat mission scoreboard, while the next plane in line has a similar number of bombs.

A formation of Dauntlesses from the 4th Marine Air Wing, returning to base after a raid on Maloelap Atoll on June 10, 1944, includes an SBD-6 numbered "14." Note the darker paint on the wingtip panel on this Dauntless. *National Museum of Naval Aviation*

A very heavily weathered SBD-6 from VMSB-231 is flying above the Southern California desert in 1944. There is a large number of bombs signifying combat missions on the side of the fuselage adjacent to the cockpit, and the national insignia has almost completely worn off the fuselage. *National Museum of Naval Aviation*

For many years, hanging above the Sea-Air Operations gallery in the National Air and Space Museum was one of two surviving examples of the SBD-6, BuNo 54605. This aircraft was accepted by the Navy on March 30, 1944. It was stricken from the Navy list on June 30, 1948, and set aside for display at the Smithsonian, likely the last SBD in Navy service.

Beneath the cowl of the SBD-6 is a Wright R-1820-66 nine-cylinder radial engine. At 1,350 horsepower, the R-1820-66 was the most powerful engine installed in a Dauntless.

A single exhaust outlet low on either side of the cowl expelled the gases from the engine. Just behind the cowling, beneath the fuselage was the oil cooler air scoop, shown here in the open position.

Unlike many other aircraft, which adopted frameless canopies during the war, the Dauntless used a heavy-frame canopy throughout production.

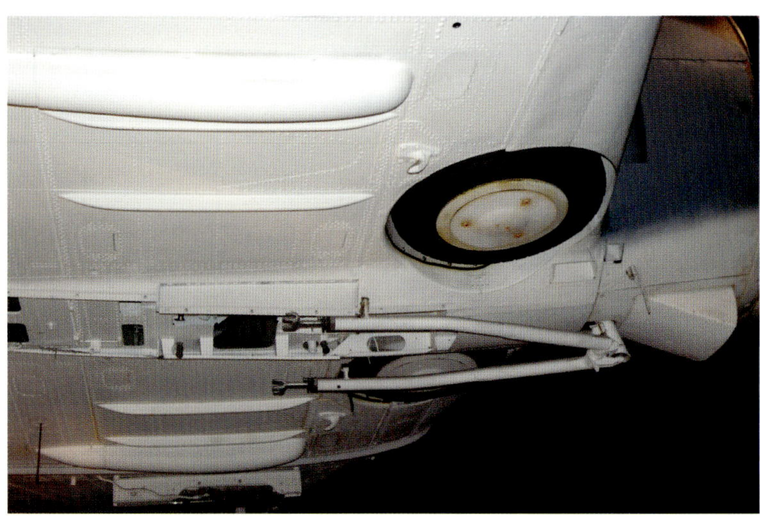

The trapeze apparatus used to swing the centerline bomb clear of the propeller arc is clearly seen on the Smithsonian's SBD-6.

Similarly, the landing gear, while always retractable, never had doors covering the main gear tires and wheels, leaving them exposed even when the landing gear was up.

The SBD-6 was the only Dauntless variant to leave the assembly line equipped with ASV radar. One of the Yagi antennas is shown here, beneath the right-side wing.

Rear armament of the SBD-6 was consistent with that of prior models, a pair of .30-caliber machine guns.

The Smithsonian's aircraft is the sixth SBD-6 built. It was used for flight testing from October 1944 through April 1948.

	Specifications					
	SBD-1	**SBD-2**	**SBD-3**	**SBD-4**	**SBD-5**	**SBD-6**
Armament	1 x .30 cal. + 2 x .50 cal.	1 x .30 cal. + 2 x .50 cal.	2 x .30 cal. + 2 x .50 cal.	2 x .30 cal. + 2 x .50 cal.	2 x .30 cal. + 2 x .50 cal.	2 x .30 cal. + 2 x .50 cal.
Bombload	1,200 lbs.	1,200 lbs.	1,200 lbs.	1,200 lbs.	1,200 lbs.	1,200 lbs.
Engine	Wright R-1820-32; 1,000 hp	Wright R-1820-32; 1,000 hp	Wright R-1820-52; 1,000 hp	Wright R-1820-52; 1,000 hp	Wright R-1820-60; 1,200 hp	Wright R-1820-66; 1,350 hp
Maximum speed	253 mph	252 mph	250 mph	245 mph	252 mph	262 mph
Service ceiling	29,600 ft.	26,000 ft.	27,100 ft.	26,700 ft.	24,300 ft.	28,600 ft.
Range	1,165 miles	1,370 miles	1,580 miles	1,450 miles	1,565 miles	1,700 miles
Wingspan	41 ft. 6 in.	41 ft. 6 in.	41 ft. 6 in.	41 ft. 6 in.	41 ft. 6 in.	41 ft. 6 in.
Length	33 ft.	33 ft.	33 ft.	33 ft.	33 ft.	33 ft.
Height	13 ft. 7 in.	13 ft. 7 in.	13 ft. 7 in.	13 ft. 7 in.	13 ft. 7 in.	13 ft. 7 in.
Weight (maximum)	9,790 lbs.	10,360 lbs.	10,400 lbs.	10,480 lbs.	10,700 lbs.	10,882 lbs.
Number built/ converted	57	87	584	780	2,965	450

The SBD-6 was the last model of Douglas Dauntless built, with the final example leaving the assembly line on July 22, 1944. Only 450 of this model were produced, bringing to a conclusion the production of arguably America's finest dive-bomber. *Stan Piet collection*